NMR SPECTROSCOPY

Basic Principles and Applications

HARCOURT BRACE JOVANOVICH COLLEGE OUTLINE SERIES

NMR SPECTROSCOPY

Basic Principles and Applications

Roger S. Macomber

Department of Chemistry
University of Cincinnati
Cincinnati, Ohio

Books for Professionals
Harcourt Brace Jovanovich, Publishers
San Diego   New York   London

Requests for permission to make copies of any part of the work should be mailed to:

Permissions
Harcourt Brace Jovanovich, Publishers
Orlando, Florida 32887

Printed in the United States of America

Library of Congress Cataloging-in-Publication Data

Macomber, Roger S.
   NMR spectroscopy: basic principles and applications. (Harcourt Brace Jovanovich college outline series) (Books for professionals)
   Includes index.
   1. Nuclear magnetic resonance spectroscopy. I. Title. II. Series. III. Series: Books for professionals.
QD96.N8M323 1988   538′.362   87-21141
ISBN 0-15-601650-8

First edition

A B C D E

PREFACE

What, precisely, is **nmr**? This book is my attempt to answer that question for anyone who wants to know. That is, if you've had a college chemistry course, and perhaps a little high school physics, you'll be able to read this book and answer that question for yourself.

Nuclear magnetic resonance (nmr) spectroscopy, originally a laboratory curiosity belonging to physicists, was adopted by chemists some time ago. In fact, nmr has developed into the single most important technique that chemists use to investigate molecular structure. And during the last few years, the technique has generated immense interest in the field of medicine. In many cases, nmr can provide better diagnostic information than that given by a CAT scan, without the deleterious effects of X-ray exposure. (In the hospital environment, however, the technique is referred to as "magnetic resonance imaging" or MRI. The term "nuclear" has been quietly deleted, presumably to still the uneasy quiver evoked by that mushroom-clouded word.) And now the usefulness of nmr is rapidly spreading to other fields — biology, geology, anthropology, forensic medicine. . . .

As popular as it is, nmr is the subject of many excellent books, some of which I can recommend highly. But most of these books are aimed at *chemists*, especially those in their upper division or postgraduate years. In writing *this* book, it has been my goal to make the topic understandable to virtually *anyone* with an interest in the field (pun intended). Chemist that I am, I naturally hope this book helps students and technicians in chemistry; but I also expect it will be useful to biology students, geology students, medical students, and even physicians, who must always be students of new technology. This book begins at a very fundamental level (electromagnetic waves), then proceeds to introduce each new topic on the basis of what came before. Upon reaching the end of the book, Special Topics, you will have been introduced to even the most sophisticated modern aspects of nmr. And, because nmr is inextricably connected with molecular structure, you'll also have picked up (or been reminded of) certain aspects of chemical structure and bonding that I've interjected at strategically appropriate places. But this is not a chemistry text, so I promise — you'll only have to cope with as much structural chemistry as is needed to explain the topic at hand.

Above all, I've tried to make the text clear, logical, and interesting (maybe even fun) to read. There are lots of illustrations, and many analogies are offered to clarify the more esoteric points. But, because of the way I structured the book, I recommend that you proceed straight through it, chapter by chapter, rather than skipping around. Still, if you *must* skip around, even peek at the end, go right ahead. I have tried to anticipate the skippers and peekers by adding liberal references to prior sections that have supporting information.

Soon after starting Chapter 1, you'll see that I have adopted a "semi-programmed" approach. That is, there are frequent example problems to test your mastery of the material covered to that point. These examples are there for a purpose. To get the most from this book, you should keep a pencil handy and try to work each problem *before* you look at the answer. But whatever you do, *don't skip the examples*, for they often contain important additional information about the topic at hand. In effect, the examples help to advance the plot. There are also two sets of inclusive review problems (after Chapters 7 and 12) which I've called "Self Tests." These problems may not advance the plot, but they'll help keep you in the picture.

It's my hope that, by mastering the material in this book, you'll be conversant with the theories, practice, and applications of nmr in today's world. It's also my hope that this book be as clear and easy to read as possible. To that end, I would be most appreciative of any feedback you might have about the book. Enjoy!

Cincinnati, Ohio   ROGER S. MACOMBER

ACKNOWLEDGMENTS

This book would never have become more than a concept if it hadn't been for the support of many individuals. It is my pleasure to acknowledge their contributions.

Mr. Philip Unitt, the Production Editor for Books for Professionals (BFP) at Harcourt Brace Jovanovich (HBJ), served tirelessly to provide a book that was as clear, logical, and error-free as possible. Ms. Emily Thompson, the Executive Editor of BFP in charge of this project, gave me the initial and continuing encouragement to present the subject of nuclear magnetic resonance in this unique way. Her unqualified support and wonderful sense of humor added greatly to my enjoyment of this endeavor. And it was only through the efforts of her husband Tom Thompson, Managing Editor of the HBJ College Department, that she became aware of my interest in such a project. Special thanks are also due to Dr. Stuart Rosenfeld, Professor of Chemistry at Smith College. In his dual capacity as Editorial Advisor to BFP and as reviewer, he read every word of the galleys and offered many useful suggestions.

Several of my colleagues at the University of Cincinnati's Department of Chemistry were most helpful in various ways. Professors John Alexander, Fred Kaplan, Albert Bobst, Jerry Ackerman (now at Massachusetts General Hospital's Department of Radiology), and Mr. Scott Smith each read sections of the manuscript and gave me their constructive criticisms. Professors R. Marshall Wison and George Kreishman, Dr. Elwood Brooks, and Dr. Alvan Hengge helped by generating several of the spectra used in this book. And Professors Estel Sprague and Tom Ridgway shared their computer graphics expertise in the production of several of the figures. I am also grateful to the students in this year's graduate course in organic spectroscopy—especially Ms. Barbara Ritshel—who served as "guinea pigs" for the first draft of this book: Their contribution was invaluable.

Dr. David Lankin, a graduate of this Department now with Varian Associates, provided me with a fruitful compilation of nmr references during the initial stages of this project. And James Shoolery of Varian, whose name is a household word to those who move in nmr circles, was most helpful with his criticisms of several chapters and with the reference materials he graciously shared. Ms. Margaret Macomber read parts of both the manuscript and galley proofs, and offered constructive comments.

It would have been impossible to write a useful book on this topic without including a large number of actual nmr spectra. I owe a debt of gratitude to Dr. Alfred Bader and Mr. Charles Pouchert of Aldrich Chemical Company and to Ms. Marie Scandone of Sadtler Research Laboratories for allowing me to reproduce spectra from *The Aldrich Library of NMR Spectra*, Sadtler's *Guide to Carbon-13 NMR Spectra*, and Sadtler's *Standard Proton NMR Spectra Collection*.

I also owe thanks to Ms. Glenda Enzweiller and Ms. Jeri Smith, for their efforts at the word processor, translating my henscratch handwriting into a legible manuscript, then putting up with endless revisions.

And finally, to my many other students who've helped me find the most successful ways to introduce this topic . . . Thank you!

Roger A Macomber

CONTENTS

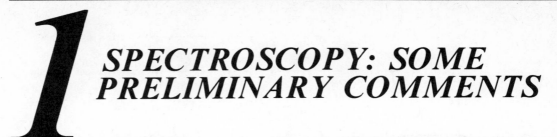

1 SPECTROSCOPY: SOME PRELIMINARY COMMENTS

THIS CHAPTER IS ABOUT

☑ **What Is NMR?**
☑ **Properties of Electromagnetic Radiation**
☑ **Interaction of Radiation with Matter**
☑ **Uncertainty and the Question of Time Scale**

1-1. What Is NMR?

Nuclear magnetic resonance (nmr) spectroscopy is the study of molecular structure by means of the interaction of radio-frequency (rf) electromagnetic radiation with a collection of nuclei immersed in a strong magnetic field. These nuclei are parts of atoms, which, in turn, are assembled into molecules. An nmr spectrum, therefore, can provide detailed information about molecular structure, information that would be difficult, if not impossible, to obtain by any other method.

It was around the turn of the century when a physicist named Zeeman first discovered that certain specific nuclei behave strangely when subjected to strong external magnetic fields. But it was not until the late 1950s that chemists began putting the so-called **Zeeman effect** to practical use. It would probably be an understatement to say that, during the succeeding years, nmr has completely revolutionized the study of chemistry. Nmr has without doubt become the single most widely used technique for elucidation of molecular structure. But before we can begin our foray into nmr, we need to review certain fundamental principles.

1-2. Properties of Electromagnetic Radiation

Because all types of spectroscopic techniques involve the interaction of matter with **electromagnetic radiation**, we should begin with a description of the properties of such radiation. The light rays that allow your eyes to see this page actually constitute electromagnetic radiation in the visible region. Each ray can be pictured as a sine wave resembling the one shown in Figure 1-1. Notice that "the" wave is actually two mutually perpendicular waves that are exactly in phase. That is, they both reach peaks, nodes, and troughs at the same points. One of these waves describes an electric field (**E**) oscillating in one plane (e.g., the plane of the page); the other describes a magnetic field (**B**) oscillating in a plane perpendicular to the electric field. The axis along which the wave propagates (the abscissa in Figure 1-1) can have dimensions of either time or length.

The wave pictured in Figure 1-1 can be characterized by two independent quantities, **wavelength** (λ) and maximum **amplitude** (E_0 and B_0 in the figure). Knowing that electromagnetic radiation travels with a fixed velocity c (3.00×10^{10} cm s^{-1} in a vacuum), we can alternatively describe the wave as having a **frequency** v, which is the inverse of the peak-to-peak time t_0 in the figure:

FREQUENCY OF A WAVE $$v = \frac{1}{t_0}$$ (1-1)

where t_0 is measured in seconds and v has units of reciprocal seconds.

note: The unit of frequency, the cycle per second (cps or s^{-1}), is called the **hertz** (Hz) in honor of the famous physicist Heinrich R. Hertz.

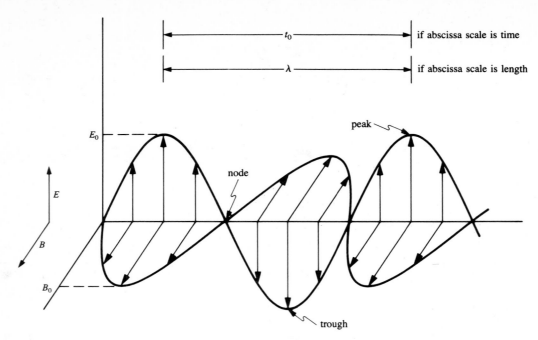

FIGURE 1-1 An electromagnetic wave.

Recognizing that the wave travels a distance λ in time t_0, we can derive our second relationship:

WAVELENGTH–FREQUENCY RELATIONSHIP
$$c = \frac{\lambda}{t_0} = \lambda v \qquad (1\text{-}2)$$

Thus, the wavelength and frequency of a given wave are not independent quantities; rather, they are inversely proportional. Radiation of high frequency has a short wavelength; conversely, radiation of low frequency has a long wavelength.

Although in principle the frequency and wavelength of electromagnetic radiation range from infinitesimal to infinite, the known **electromagnetic spectrum** ranges from cosmic rays of extremely high frequency (and short wavelength) to rf radiation of low frequency (and long wavelength). The various regions of the electromagnetic spectrum are listed in Table 1-1. You'll note, for example, that the visible region comprises radiation of wavelength 380–780 nm (1 nm = 10^{-9} m = 10^{-7} cm) and frequency $4-8 \times 10^{14}$ Hz. Our optic nerves do not respond to "light" outside this region.

In addition to its wave-like properties, electromagnetic radiation also exhibits certain behavior characteristic of particles. Such a particle, or **quantum**, of radiation is called a **photon**. For our purposes, the most important particle-like property is *energy*. Each photon possesses a discrete amount of energy

TABLE 1-1 The Electromagnetic Spectrum

Radiation	λ Wavelength (nm)	v Frequency (Hz)	Energy (kJ mol^{-1})
Cosmic rays	$<10^{-3}$	$>3 \times 10^{20}$	$>1.2 \times 10^8$
Gamma rays	10^{-1} to 10^{-3}	3×10^{18} to 3×10^{20}	1.2×10^6 to 1.2×10^8
X-rays	10 to 10^{-1}	3×10^{16} to 3×10^{18}	1.2×10^4 to 1.2×10^6
Far ultraviolet	200 to 10	1.5×10^{15} to 3×10^{16}	6×10^2 to 1.2×10^4
Ultraviolet	380 to 200	8×10^{14} to 1.5×10^{15}	3.2×10^2 to 6×10^2
Visible	780 to 380	4×10^{14} to 8×10^{14}	1.6×10^2 to 3.2×10^2
Infrared	3×10^4 to 780	10^{13} to 4×10^{14}	4 to 1.6×10^2
Far infrared	3×10^5 to 3×10^4	10^{12} to 10^{13}	0.4 to 4
Microwaves	3×10^7 to 3×10^5	10^{10} to 10^{12}	4×10^{-3} to 0.4
Radio frequency (rf)	10^{11} to 3×10^7	10^6 to 10^{10}	4×10^{-7} to 4×10^{-3}

that is directly proportional to its frequency (if we regard it as a wave). This relationship can be written

ENERGY OF A PHOTON $E_{photon} = h\nu_{photon}$ **(1-3)**

where h, **Planck's constant**, has values of 6.63×10^{-34} J s per photon. Alternatively, h can be expressed on a "per mole" basis through multiplication by Avogadro's number (6.02×10^{23} mol^{-1}) and division by 10^3 J kJ^{-1} to give $h = 3.99 \times 10^{-13}$ kJ s mol^{-1}. Since the strength of a typical chemical bond is around 400 kJ mol^{-1}, radiation above the visible region has sufficient energy to break chemical bonds, while radiation below the visible region does not (see Table 1-1). Of particular interest to us for nmr is **radio-frequency radiation**, the same type that carries communication signals to our radios and televisions. Such radiation, with frequencies on the order of 100 MHz (1 MHz = 10^6 Hz), is at the lowest end of the energy scale. This, it will turn out, is exactly the amount of energy we will need to perform nmr experiments.

EXAMPLE 1-1 Derive the relationship between the energy of a photon and its wavelength.

Solution: We can rearrange Eq. (1-2) to

$$\nu = \frac{c}{\lambda}$$

Substituting for ν in Eq. (1-3) gives us

$$E_{photon} = h\nu_{photon} = \frac{hc}{\lambda_{photon}}$$

1-3. Interaction of Radiation with Matter: Quantum Mechanics Rears Its Ugly Head

Now that we know something about electromagnetic radiation, let's turn to the question of what factors control the interaction of such radiation with particles of matter. **Quantum mechanics**, the arcane field of physics that deals with energy at the microscopic (atomic) level, tells us there is one consistent requirement shared by all forms of absorption spectroscopy: For a particle to absorb a photon of electromagnetic radiation, the particle must first be in some sort of uniform *periodic* motion. Furthermore, the frequency of that motion must *exactly* match the frequency of absorbed radiation:

FREQUENCY MATCHING $\nu_{motion} = \nu_{photon}$ **(1-4)**

This fact, which might initially appear to be an incredible coincidence, is actually quite logical. If the photon is to be absorbed, its energy, which is originally in the form of the oscillating electric and magnetic fields, must be transformed into energy of the particle's motion. This transfer of energy can take place only if the oscillations of the electric and/or magnetic field can constructively interfere with the "oscillations" (uniform periodic motion) of the particle. When such a condition exists, the system is said to be in **resonance**, and only then can the act of absorption take place. Don't confuse the term resonance in this context with the resonance (conjugation) interactions of electrons within molecules.

EXAMPLE 1-2 The C=O bond in formaldehyde vibrates (stretches, then contracts) with a frequency of 5.13×10^{13} Hz. **(a)** What frequency of radiation could be absorbed by this vibrating bond? **(b)** How much energy would each photon deliver? **(c)** To which spectral region does this radiation belong?

Solution:

(a) From Eq. (1-4) we know the frequencies must match; therefore, $\nu_{photon} = 5.13 \times 10^{13}$ Hz.
(b) From Eq. (1-3),

$$E_{photon} = h\nu = (6.63 \times 10^{-34} \text{ Js})(5.13 \times 10^{13} \text{ s}^{-1}) = 3.40 \times 10^{-20} \text{ J} = 20.5 \text{ kJ mol}^{-1}$$

(c) From Table 1-1 we see that radiation of this frequency and energy falls in the infrared region.

At this point you might think that the frequency-matching requirement places a heavy constraint on the types of absorption processes that can occur. After all, how many kinds of periodic motion can a particle have? The answer is that even a small molecule is constantly undergoing many types of periodic motion. Each of its bonds is constantly vibrating; the molecule as a whole and all its individual parts are rotating in all three dimensions; the electrons are circulating through their orbitals. And each of these processes has its own characteristic frequency!

All of these forms of microscopic motion are *inherent*. That is, the motion takes place all by itself, without intervention by any external agent. However, it should be possible in certain circumstances to *induce* the particles to engage in some additional form of periodic motion. Then, to achieve resonance, all we need to do is match the frequency of this induced motion with that of the incident radiation [Eq. (1-4)]. For example, an ion (or any charged particle, for that matter) follows a curved path as it moves through a magnetic field. If we carefully adjust the strength of the magnetic field for the ion's mass and charge, the ion will follow an exactly circular path with a characteristic fixed frequency. Matching this characteristic frequency with incident radiation of the same frequency is the basis of a technique known as **ion cyclotron resonance spectroscopy**. We'll discover in Chapter 2 that a strong magnetic field can also be used to induce certain nuclei to move with uniform periodic motion of a different type.

1-4. Uncertainty and the Question of Time Scale

If you've ever tried to take a photograph of a moving object, you know that the shutter speed of the camera must be adjusted to avoid blurring the image. And, of course, the faster the object is moving, the shorter must be the exposure time to "freeze" the motion. We have very similar considerations in spectroscopy.

Suppose you owned a collection of very extraordinary chameleons that were able to change colors from blue to green and back again in a small fraction of a second. If you took a picture of them with a shutter speed of one second, each of the little critters would appear to be turquoise. But if you decreased the exposure time to 0.01 second, the photograph would show blue ones and green ones in roughly equal numbers, and no turquoise ones! Thus, to capture the individual colors, your exposure time must be shorter than the time required for the color change.

There are many types of molecular chameleons, i.e., molecules that are constantly undergoing some sort of reversible reorganization of their atoms and bonds. If absorption of the photon is fast enough, we will detect both the "blue" and "green" forms of the molecule. But if the absorption process is slower than the interconversion, we will detect only some sort of "average" image. The situation, therefore, boils down to the question: How long does it take for a particle to absorb a photon? Unfortunately, such a question is impossible to answer with complete accuracy.

In 1927, Heisenberg, one of the early practitioners of quantum mechanics, stated his now-famous **uncertainty principle**: There will always be a limit to the accuracy with which we can *simultaneously* determine the energy and time scale of an event. Mathematically, the product of the uncertainties can never be less than h, our old friend, Planck's constant:

UNCERTAINTY PRINCIPLE $$\Delta E \, \Delta t \geq h \tag{1-5}$$

Thus, if we knew the frequency of a given photon to a high order of accuracy, we would be unable to measure accurately how long the photon takes to be absorbed. Nonetheless, there is a useful generalization we *can* make. The time required for a photon to be absorbed must be *at least* as long as it takes one full "cycle" of the wave to pass the particle. This length of time, which we called t_0 in Figure 1-1, is nothing more than $1/v$. This stands to reason if we consider that the particle would have to wait through at least one full cycle before it could sense what the radiation frequency was. At least we now have an order-of-magnitude idea of how fast our shutter speed must be.

EXAMPLE 1-3 Suppose our nmr experiment required the use of rf radiation, with a frequency of 100 MHz, to examine formaldehyde (see Example 1-2). Will the experiment enable us to see the various lengths of the C=O bond as it stretches and contracts, or will it see only an average bond length?

Solution: The vibrational time scale ($1/v = 1.9 \times 10^{-14}$ s) is much shorter (faster) than the nmr time scale ($1/v = 1 \times 10^{-8}$ s). Therefore, only an "average" $C{=}O$ bond length could be detected by nmr.

Equipped with our new knowledge about electromagnetic radiation, periodic motion, resonance, and time scale, we're now ready to enter the intriguing world of the atomic nucleus.

SUMMARY

1. Nuclear magnetic resonance (nmr) spectroscopy involves the interaction of certain nuclei with radio-frequency (rf) electromagnetic radiation when the nuclei are immersed in a strong magnetic field.
2. Electromagnetic radiation is characterized by its frequency (v) or wavelength (λ), which are inversely proportional:

$$v\lambda = c \qquad \text{(1-2)}$$

where c is the speed of light. Rf radiation has frequencies on the order of 100 MHz.
3. The energy of a photon (or quantum of radiation) is given by

$$E = hv \qquad \text{(1-3)}$$

where h is Planck's constant.
4. For radiation to be absorbed by a particle, the frequency of the radiation must equal the frequency of the particle's periodic motion.
5. The Heisenberg Uncertainty Principle [Eq. (1-5)] defines the time scale of radiation absorption event as the inverse of the radiation's frequency. Processes that occur faster than the spectroscopic time scale are "averaged" during the absorption process.

2 MAGNETIC PROPERTIES OF NUCLEI

THIS CHAPTER IS ABOUT

☑ **The Structure of an Atom**
☑ **The Nucleus in a Magnetic Field**
☑ **Nuclear Energy Levels and Relaxation**

2-1. The Structure of an Atom

The compounds we will examine by nmr are composed of molecules, which are themselves aggregates of atoms. Each atom has some number of **electrons** circulating around a tiny, dense bit of matter called the **nucleus**. Virtually all of the mass of an atom is concentrated in its nucleus, although the nucleus occupies only $1/10^{12}$ of the atom's volume. Even the nucleus can be further dissected into more fundamental particles including **protons** and **neutrons**, not to mention a host of other subnuclear particles that help hold the nucleus together and give nuclear physicists something to wonder about.

A. The composition of the nucleus

It is the number of protons in its nucleus (Z, the atomic number) that determines both an atom's identity and the charge on its nucleus. Every nucleus with just one proton is a hydrogen nucleus, every nucleus with six protons is a carbon nucleus, and so on. Yet, if we carefully examine a large sample of hydrogen atoms, we find their nuclei are not all identical. It's true that all have just one proton, but they differ in the number of neutrons. Most hydrogen atoms in nature (99.985% of them, to be exact) have no neutrons ($N = 0$), but a small fraction (0.015%) have one neutron ($N = 1$) in each nucleus. These two forms are the naturally occurring **isotopes** of hydrogen, and they are given the symbols ^1H and ^2H, respectively. The superscript refers to the **nominal atomic mass** (A) of the atom, which is the sum of Z and N.

NOMINAL ATOMIC MASS $$A = Z + N \qquad (2\text{-}1)$$

^2H is usually referred to as deuterium or heavy hydrogen, but most isotopes of other elements are identified simply by their value of A. Note that the atomic mass listed for each element in the periodic table (Appendix 1) is a weighted average over all naturally occurring isotopes.

EXAMPLE 2-1 Tritium, ^3H, is a radioactive (unstable) isotope of hydrogen. What is the composition of its nucleus?

Solution: Since the atom is an isotope of hydrogen, $Z = 1$. A is 3 and therefore, from Eq. (2-1), $N = 2$. Thus, the nucleus consists of one proton and two neutrons.

EXAMPLE 2-2 Natural chlorine ($Z = 17$) is composed of two isotopes, ^{35}Cl and ^{37}Cl. The atomic mass listed for chlorine in the periodic table is 35.5. (a) What is the composition of each nucleus? (b) What is the natural abundance of each isotope?

Solution:

(a) ^{35}Cl has $Z = 17$ (17 protons), $A = 35$, and $N = 18$ (18 neutrons); ^{37}Cl has $Z = 17$, $A = 37$, and $N = 20$.

(b) Since the atomic mass of 35.5 is a weighted average of a mixture of ^{35}Cl and ^{37}Cl, we can use algebra to calculate the fraction (f) of each isotope:

$$(f_{35} \cdot 35) + (f_{37} \cdot 37) = 35.5$$

and

$$f_{35} + f_{37} = 1.00$$

Therefore

$$(f_{35} \cdot 35) + (1.00 - f_{35})(37) = 35.5$$

$$f_{35} = 0.75 \ (75\%)$$

$$f_{37} = 0.25 \ (25\%)$$

B. Electron spin

Before we delve further into the properties of the nucleus, let's momentarily shift our attention back to one of the electrons zooming around the nucleus. Just like photons, electrons exhibit both wave and particle properties. Electron waves are characterized by four **quantum numbers**. The first three of these describe the energy, shape, and orientation of the volume the electron occupies in the atom. This volume is called an **orbital**. The fourth quantum number is the **electron spin quantum number** s, which can assume only two values, $+\frac{1}{2}$ or $-\frac{1}{2}$. (Why $\pm\frac{1}{2}$ was selected rather than, say, ± 1 is described a little later.) The **Pauli exclusion principle** tells us that if two electrons occupy the *same* orbital, they must have *different* spin quantum numbers. Therefore, no orbital can possess more than two electrons, and then only if their spins are *paired* (opposite).

But, is there any other significance to the spin quantum number? Yes, indeed! Because it can be regarded as a particle spinning on its axis, the electron has a property called **spin angular momentum**. Further, because the electron is a *charged* particle ($Z = -1$), its spinning gives rise to a tiny magnetic moment represented by the boldface vector arrows in Figure 2-1. The two possible values of s correspond to the two possible orientations of the magnetic moment vector, up or down. (Why there are only two possible orientations, rather than any other number, is simply a fact of life for which no better explanation yet exists.) These two **spin states** of the electron are *degenerate* (i.e., have the same energy) in the absence of an external magnetic field. However, when the electrons *are* immersed in such a field, the two states are no longer degenerate. An electron oriented antiparallel (opposite) to the field ($s = -\frac{1}{2}$ in Figure 2-1) has lower energy than an electron parallel to the field ($s = +\frac{1}{2}$). This fact is centrally important in the technique known as **electron paramagnetic resonance spectroscopy** (Chapter 10). But now, let's return to the nucleus.

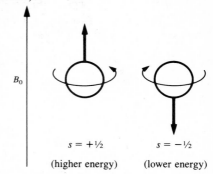

$s = +\frac{1}{2}$ $s = -\frac{1}{2}$

(higher energy)   (lower energy)

FIGURE 2-1 Spin angular momentum of the electron. B_0 represents the orientation of the external magnetic field.

C. Nuclear spin

The proton is a spinning charged ($Z = 1$) particle too, so it shouldn't surprise us to learn that it exhibits a magnetic moment also. And, like the electron's, its magnetic moment has only two possible orientations, which are degenerate in the absence of an external magnetic field. To differentiate **nuclear spin states** from electronic spin states, we will adopt the usual convention of labeling nuclear spin states with the **nuclear spin quantum number** m. Thus, for a proton, m can assume values of only $+\frac{1}{2}$ or $-\frac{1}{2}$. We describe such a nucleus as having a **nuclear spin** (I) of $\frac{1}{2}$. But because nuclear charge is the opposite of electron charge, a nucleus whose magnetic moment is *parallel* to the magnetic field has the lower energy (Figure 2-2).

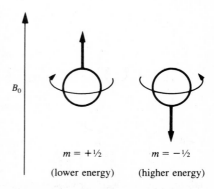

$m = +\frac{1}{2}$ $m = -\frac{1}{2}$

(lower energy)   (higher energy)

FIGURE 2-2 Spin angular momentum of the proton. B_0 represents the orientation of the external magnetic field.

So, we have established that ^{1}H nuclei (protons) exhibit two possible magnetic spin orientations. What about other isotopes? From Chapter 1 you might remember that Zeeman found that only certain nuclei respond to an external magnetic field. This is because only certain isotopes possess nonzero nuclear spin. Here are some rules that allow us to predict whether or not a given isotope has this all-important property:

1. A nucleus with an *even* mass A and *even* charge Z, and therefore also an *even* N, will have a nuclear spin I of zero. Examples include ^{12}C, ^{16}O, ^{18}O, and ^{32}S.

   • Nuclei with $I = 0$ lack magnetic properties and cannot be detected by nmr.

2. A nucleus with *even* mass and *odd* charge (both Z and N odd) will exhibit an integer value of I and *will* be detectable by nmr. Examples are ^{2}H ($I = 1$), ^{10}B ($I = 3$), ^{14}N ($I = 1$), and ^{50}V ($I = 6$).

3. A nucleus with *odd* mass (Z odd and N even, or Z even and N odd) will have nuclear spin with and I value that we can express as $n/2$, where n is an odd integer. Here are some examples: ^{1}H ($I = \frac{1}{2}$), ^{11}B ($I = \frac{3}{2}$), ^{13}C ($I = \frac{1}{2}$), ^{15}N ($I = \frac{1}{2}$), ^{17}O ($I = \frac{5}{2}$), ^{19}F ($I = \frac{1}{2}$), ^{29}Si ($I = \frac{1}{2}$), ^{31}P ($I = \frac{1}{2}$). These nuclei, too, can be detected by nmr. Notice also that

   • Different isotopes of the same element have different nuclear spins, some of which *are* detectable by nmr, others of which are *not*.

EXAMPLE 2-3 Predict the nuclear spin I of ^{4}He, ^{6}Li, and ^{7}Li, and indicate which are detectable by nmr.

Solution: Assign each nucleus to one of the three classes described above.

Nucleus	A	Z	N	Class	Predicted I	Detectable?
^{4}He	4	2	2	1	0	no
^{6}Li	6	3	3	2	integer	yes
^{7}Li	7	3	4	3	$n/2$	yes

The actual values of I are ^{4}He, $I = 0$; ^{6}Li, $I = 1$; and ^{7}Li, $I = \frac{3}{2}$.

Just when we begin to understand nuclear spin numbers of $\frac{1}{2}$, we are confronted with values of $\frac{3}{2}$, $\frac{5}{2}$, and even 6. What is the meaning of this? The explanation is actually quite straightforward. Although *single* atomic particles such as protons and electrons can adopt only two possible magnetic spin orientations, more complex nuclei can adopt more than two. Furthermore, the total number of possible spin states (values of m) is determined directly by the value of I.

NUMBER OF POSSIBLE SPIN STATES   number of states $= 2I + 1$ (2-2)

Each of these states has its own spin quantum number m in the range $m = -I, -I + 1, \ldots, I - 1, I$. Thus, for nuclei with $I = \frac{1}{2}$, only two states are possible: $m = +\frac{1}{2}$ and $m = -\frac{1}{2}$.

EXAMPLE 2-4 How many spin states are possible for each of these nuclei, and what is the value of m for each state?

$$^{11}\text{B}, \quad ^{12}\text{C}, \quad ^{14}\text{N}, \quad ^{17}\text{O}, \quad ^{31}\text{P}$$

Solution: The I values for these nuclei are listed above. Using Eq. (2-2), we can find first the number of spin states and then their m values.

Nucleus	I	Number of states	m values
^{11}B	$\frac{3}{2}$	4	$-\frac{3}{2}, -\frac{1}{2}, \frac{1}{2}, \frac{3}{2}$
^{12}C	0	1	0
^{14}N	1	3	$-1, 0, 1$
^{17}O	$\frac{5}{2}$	6	$-\frac{5}{2}, -\frac{3}{2}, -\frac{1}{2}, \frac{1}{2}, \frac{3}{2}, \frac{5}{2}$
^{31}P	$\frac{1}{2}$	2	$-\frac{1}{2}, \frac{1}{2}$

2-2. The Nucleus in a Magnetic Field

A. Nuclear Zeeman effect

As we have said, a nucleus with spin I can adopt $2I + 1$ spin orientations, which are degenerate outside a magnetic field. However, when a collection of these nuclei is immersed in a magnetic field, the spin states are no longer degenerate. They separate in energy, with the largest positive m value corresponding to the lowest-energy (most stable) state. It is this separation of states in a magnetic field that is referred to as the **nuclear Zeeman effect**.

The energy of a given spin state (E_i) is directly proportional to the value of m_i and the magnetic field strength B_0.

SPIN STATE ENERGY
$$E_i = -m_i B_0 \frac{\gamma h}{2\pi} \tag{2-3}$$

In this equation h (Planck's constant) and π have their usual meanings, while γ is called the **magnetogyric ratio**, a constant characteristic of the nucleus being examined (more on this a little later). The minus sign follows from the convention of making a positive m correspond to a lower (negative) energy. Figure 2-3 graphically depicts the variation of spin state energy versus magnetic field strength for two different nuclei, one with $I = \frac{1}{2}$, the other with $I = 1$. Notice that as field strength increases, the *difference* in energy between any two spin states increases proportionally.

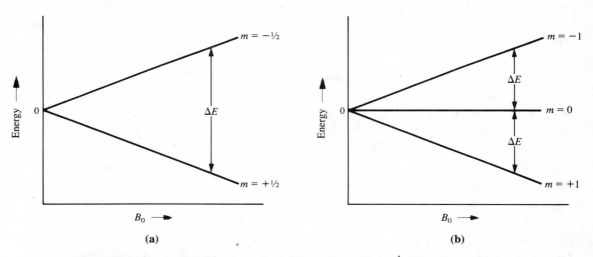

FIGURE 2-3 The nuclear Zeeman effect. (**a**) Nucleus with $I = \frac{1}{2}$; (**b**) nucleus with $I = 1$.

For a nucleus with $I = \frac{1}{2}$, this difference, ΔE, is

$$\Delta E = E_{-1/2} - E_{+1/2}$$
$$= -B_0 \left[\left(-\frac{1}{2} \right) - \left(+\frac{1}{2} \right) \right] \frac{\gamma h}{2\pi} \tag{2-4}$$
$$= \frac{\gamma h B_0}{2\pi}$$

And now you might realize why values of $\pm\frac{1}{2}$ were picked for m (and s too, for that matter). It is so the *difference* in energy between two neighboring states will always be an integer multiple of B_0 $(\gamma h/2\pi)$.

The magnetogyric ratio γ is a proportionality constant that describes the spin state energies of a given nucleus in an external magnetic field. Each isotope with nonzero nuclear spin has its own unique value of γ, though the magnitude of γ depends on the units selected for B_0. We will use the unit tesla (T) for magnetic field strength so that γ has units of $\mathrm{rad\,T^{-1}\,s^{-1}}$.

note: Earlier books on nmr used gauss (or kilogauss) for magnetic field strength;
    $1\,\mathrm{T} = 10^4$ gauss $= 10$ kilogauss.

In Table 2-1 are listed many of the common isotopes examined by nmr techniques, together with their nuclear constants. Notice that a bare proton has the largest γ value, while heavier nuclei surrounded by many subvalence electrons tend toward lower values. This will become significant later. Relative sensitivity is the strength of the nmr signal that can be obtained from a certain number of nuclei of an isotope as compared to the strength of the signal obtained from an equal number of ^{1}H nuclei. If we compare the data on natural abundance and sensitivity, we see that the most easily studied nuclei are ^{1}H, ^{19}F, and ^{31}P. Indeed, prior to the late 1960s these three were the only nuclei routinely studied with commercially available instrumentation. More recently, however, instruments have become available that can examine a wide variety of other nuclei, including ^{13}C, which is of immense importance to organic chemists.

TABLE 2-1 Nuclear Properties of Selected Isotopes[1]

Isotope	Relative natural abundance (%)	Z	N	A	I	$\gamma/10^6$ $(\mathrm{rad\,T^{-1}\,s^{-1}})$	ν (MHz at 1 T)	Relative sensitivity[2]
^{1}H	99.985	1	0	1	$\frac{1}{2}$	267.512	42.5759	1.00
^{2}H	0.015	1	1	2	$\frac{1}{2}$	41.0648	6.53566	9.65×10^{-3}
^{7}Li	92.58	3	4	7	$\frac{3}{2}$	103.96	16.546	0.293
^{10}B	19.58	5	5	10	3	28.748	4.5754	1.99×10^{-2}
^{11}B	80.42	5	6	11	$\frac{3}{2}$	85.828	13.660	0.165
^{13}C	1.108	6	7	13	$\frac{1}{2}$	67.2640	10.7054	1.59×10^{-2}
^{14}N	99.63	7	7	14	1	19.325	3.0756	1.01×10^{-3}
^{15}N	0.37	7	8	15	$\frac{1}{2}$	27.107	4.3142	1.04×10^{-3}
^{17}O	0.037	8	9	17	$\frac{5}{2}$	36.27	5.772	2.91×10^{-2}
^{19}F	100	9	10	19	$\frac{1}{2}$	251.667	40.0541	0.833
^{23}Na	100	11	12	23	$\frac{3}{2}$	70.761	11.262	9.25×10^{-2}
^{27}Al	100	13	14	27	$\frac{5}{2}$	69.706	11.094	0.206
^{29}Si	4.70	14	15	29	$\frac{1}{2}$	53.142	8.4578	7.84×10^{-3}
^{31}P	100	15	16	31	$\frac{1}{2}$	108.29	17.235	6.63×10^{-2}
^{33}S	0.76	16	17	33	$\frac{3}{2}$	20.517	3.2654	2.26×10^{-3}
^{35}Cl	75.53	17	18	35	$\frac{3}{2}$	26.212	4.1717	4.70×10^{-3}
^{37}Cl	24.47	17	20	37	$\frac{3}{2}$	21.82	3.472	2.71×10^{-3}

[1] Abstracted in part from *The 64th CRC Handbook of Chemistry and Physics*, CRC Press, 1984.
[2] Relative (to proton) sensitivity for equal numbers of nuclei at constant field; $S = 7.652 \times 10^{-3}\,\mu^3\,(I+1)/I^2$ where μ is the magnetic moment of the nucleus.

EXAMPLE 2-5 **(a)** What is the energy difference between the two spin states of ^{1}H in a magnetic field of 2.35 T? **(b)** Of ^{13}C?

Solution:

(a) Use Eq. (2-4) and the data in Table 2-1.

$$\Delta E = \frac{\gamma h B_0}{2\pi}$$

$$= \frac{(267.512 \times 10^6 \ \text{rad} \ \text{T}^{-1}\text{s}^{-1})(6.63 \times 10^{-34} \ \text{J s})(2.35 \ \text{T})}{2(3.14 \ \text{rad})}$$

$$= 6.63 \times 10^{-26} \ \text{J}$$

(b) For ^{13}C, $\gamma = 67.2640 \times 10^6$, so $\Delta E = 1.67 \times 10^{-26}$ J, about one-fourth the difference for ^{1}H.

B. Precession and the Larmor frequency

We now know that nuclei with $I \neq 0$, when immersed in a magnetic field, adopt $2I + 1$ spin orientations, each with a different energy. But before these nuclei can absorb photons, the nuclei must be in some sort of uniform periodic motion. Interestingly enough, the motion already exists! The two spin orientations of ^{1}H nuclei, for example, are *not* aligned statically with or against the external magnetic field, as Figure 2-2 suggests; instead, the magnetic moments wobble around the axis of the applied field.

If you've ever played with a spinning top, you know that its spinning is what prevents the top from falling on its side. The wobbling motion it assumes in a *gravitational* field has a uniform frequency and is called **precession**. The Earth wobbles on its axis in much the same way, though much more slowly. In an exactly analogous way, the magnetic moment of a spinning nucleus also precesses with a characteristic angular frequency, called the **Larmor frequency** ω, which is a function of γ and B_0.

LARMOR FREQUENCY $$\omega = \gamma B_0 \qquad (2\text{-}5)$$

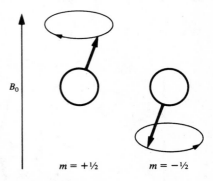

The angular Larmor frequency, in units of radians per second (rad s^{-1}), can be transformed into linear frequency v (s^{-1} or Hz) by division by 2π:

LINEAR PRECESSION FREQUENCY $$v_{\text{precession}} = \frac{\omega}{2\pi} = \frac{\gamma B_0}{2\pi} \qquad (2\text{-}6)$$

This precession of the magnetic moment is depicted in Figure 2-4. Note also that the precession frequency is *independent* of m, so that

- all spin orientations of a given nucleus precess at the same frequency in a fixed magnetic field.

FIGURE 2-4 Precession of a nuclear magnetic moment in an external magnetic field. B_0 represents the orientation of the applied field.

EXAMPLE 2-6 (a) At 2.35 T, what is the precession frequency v of an ^{1}H nucleus? A ^{13}C nucleus? (b) In what region of the electromagnetic spectrum does radiation of these frequencies occur?

Solution:

(a) For ^{1}H, using the γ value from Table 2-1, we find

$$v = \frac{\gamma B_0}{2\pi} = \frac{(267.512 \times 10^6 \ \text{rad} \ \text{T}^{-1}\text{s}^{-1})(2.35 \ \text{T})}{2(3.14 \ \text{rad})}$$

$$= 1.000 \times 10^8 \ \text{s}^{-1} = 1.000 \times 10^8 \ \text{Hz} = 100.0 \ \text{MHz}$$

Similarly, for ^{13}C,

$$v = 25.16 \ \text{MHz}$$

(b) From Table 1-1 we note that these frequencies fall in the rf (radio-frequency) region.

After you've completed this calculation the "long" way, using B_0 and γ values from Table 2-1, try an easier way. The data in the column labeled v are the precession frequency (in MHz) of each nucleus in a 1.00-T magnetic field. By simply multiplying these numbers by the actual field strength, you can readily calculate the value of v at any other field strength. Thus, for ^{1}H

$$v = (42.5759 \text{ MHz T}^{-1})(2.35 \text{ T}) = 100.0 \text{ MHz}$$

EXAMPLE 2-7 At what field strength do protons (^{1}H nuclei) precess at a frequency of 300 MHz?

Solution: Rearranging Eq. (2-6), we find

$$B_0 = \frac{2\pi v}{\gamma}$$

$$= \frac{2(3.14 \text{ rad})(300 \times 10^6 \text{ s}^{-1})}{267.512 \times 10^6 \text{ rad T}^{-1}\text{s}^{-1}}$$

$$= 7.04 \text{ T}$$

We're almost ready to perform an nmr experiment. We've immersed our collection of nuclei in a magnetic field; each is precessing with a characteristic frequency. To observe resonance, all we have to do is irradiate them with electromagnetic radiation of the appropriate frequency, right? Well, almost!

2-3. Nuclear Energy Levels and Relaxation

A. The Boltzman distribution and saturation

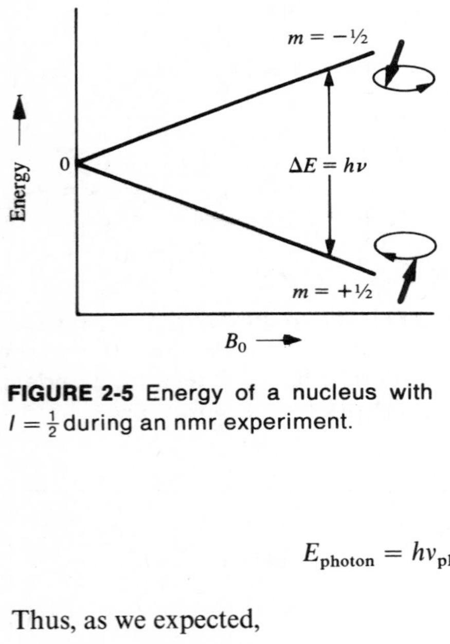

FIGURE 2-5 Energy of a nucleus with $I = \frac{1}{2}$ during an nmr experiment.

In Chapter 1 it was hinted that once a particle absorbs a photon, the energy originally associated with the electromagnetic radiation appears somehow in the particle's motion. Where does the energy go in the case of precessing ^{1}H nuclei? Because there are only two spin states possible, the energy goes into a spin "flip." That is, the photon's energy is absorbed by a nucleus in the lower-energy spin state ($m = +\frac{1}{2}$), and it is transformed into a nucleus in the higher-energy spin state ($m = -\frac{1}{2}$). This situation is depicted in Figure 2-5. And remember that this spin flip does *not* change the precessional frequency of the nucleus.

We have already calculated the energy gap between these two spin states [Eq. (2-4)] and this must equal the energy of the absorbed photon [Eq. (1-3)]. Combining these with Eq. (2-6) gives us

$$E_{\text{photon}} = hv_{\text{photon}} = \Delta E = \frac{\gamma h B_0}{2\pi} = hv_{\text{precession}} \qquad (2\text{-}7)$$

Thus, as we expected,

- For resonance to occur, the radiation frequency must exactly match the precessional frequency.

But, there is a fly in the ointment. Quantum mechanics tells us that, for net absorption of radiation to occur, there *must* be more particles in the lower-energy state than in the higher one. If the two populations happen to be equal, Einstein predicted theoretically that transition from the upper ($m = -\frac{1}{2}$) state to the lower ($m = +\frac{1}{2}$) state (a process called **stimulated emission**) is *exactly* as likely to occur as absorption. In such a case, no *net* absorption is possible, a condition called **saturation**.

Is there any reason to expect that there will be an excess of nuclei in the lower spin state? The answer is a qualified yes. For any system of energy levels *at thermal equilibrium*, there will always be more particles in the lower state(s) than in the upper state(s). However, there will still be *some* particles in the upper state(s). What we really need is an equation relating the energy gap (ΔE) between the states and the relative numbers of particles in each of those states. This time, quantum mechanics comes to our rescue in the form of the **Boltzmann distribution**.

BOLTZMANN DISTRIBUTION
$$\frac{P_{m=-1/2}}{P_{m=+1/2}} = e^{-\Delta E/kT} \tag{2-8}$$

where P is the fraction of the particle population in each state, T is the absolute temperature, and k (the Boltzmann constant) has a value of $1.381 \times 10^{-28}\ \mathrm{J\,K^{-1}}$.

EXAMPLE 2-8 (a) At 25°C (298 K) what fraction of ^{1}H nuclei in 2.35-T field are in the upper and lower states? (b) Of ^{13}C nuclei? See Example 2-5 for the values of ΔE.

Solution:

(a) Use Eq. (2-8) and the results of Example 2-5.

$$\frac{P_{-1/2}}{P_{+1/2}} = e^{-\Delta E/kT} = e^{-(6.63 \times 10^{-26}\,\mathrm{J})/(1.381 \times 10^{-23}\,\mathrm{J\,K^{-1}})(298\,\mathrm{K})}$$

$$= 0.999984$$

Since $P_{+1/2} = 1 - P_{-1/2}$,

$$\frac{P_{-1/2}}{1 - P_{-1/2}} = 0.999984 \qquad P_{-1/2} = 0.4999959 \qquad P_{+1/2} = 0.5000041$$

(b) Similarly, for ^{13}C

$$P_{-1/2} = 0.4999990 \qquad P_{+1/2} = 0.5000010$$

As you can see from the above example (you *did* do it, didn't you?), the *difference* in populations of the two states is exceedingly small, on the order of a few parts per million. And the difference for ^{13}C is even smaller than the difference for ^{1}H because of ^{13}C's smaller γ value. But don't despair. This difference *is* sufficient to generate an nmr signal. It is this small difference, however, that accounts in part for the relatively low sensitivity of nmr compared to other absorption techniques such as infrared and ultraviolet spectroscopy. Moreover, anything (such as a larger value of B_0 or sensitivity, Table 2-1) that *increases* the population difference will give rise to a more intense nmr signal.

B. Relaxation processes

Our nmr theory is almost complete, but there is one more consideration before we set about designing a spectrometer. We indicated previously that in the *absence* of an external field, all the spin states are degenerate and, therefore, are of equal probability and population. A relevant question is: How long after immersion in an external field does it take for a collection of nuclei to reach thermal equilibrium (and a Boltzmann distribution)? This process is *not* infinitely fast. In fact, the rate at which thermal equilibrium is established is governed by a quantity called the **spin–lattice** (or **longitudinal**) **relaxation time**, T_1. The exact relation is

SPIN–LATTICE RELAXATION
$$P_{eq} - P = (P_{eq} - P)_0 e^{-t/T_1} \tag{2-9}$$

where $P_{eq} - P$ is the difference between the equilibrium population and the population at time t, and the subscript zero refers to $t = 0$.

EXAMPLE 2-9 In terms of T_1, how much time is required for an initially 50/50 distribution of ^{1}H spin states to progress 95% of the way toward thermal equilibrium?

Solution: Use Eq. (2-9) with $P_{eq} - P = 0.05(P_{eq} - P)_0$ and solve for t.

$$0.05(P_{eq} - P)_0 = (P_{eq} - P)_0 e^{-t/T_1}$$

$$0.05 = e^{-t/T_1}$$

Taking the natural log (ln) of both sides of the equation, we find

$$-3.00 = -t/T_1 \quad \text{or} \quad t = 3T_1$$

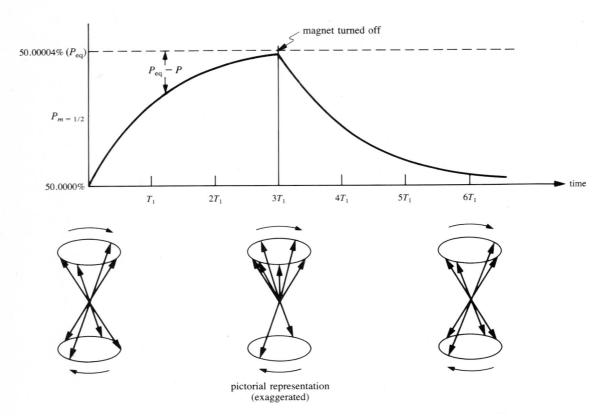

FIGURE 2-6 The exponential approach to thermal equilibrium, governed by T_1.

Figure 2-6 shows Eq. (2-9) and Example 2-9 for a typical collection of ^1H nuclei at 2.35 T. Notice from the figure that if the magnetic field is turned off, the collection decays back to the original (50/50) distribution with a rate also governed by T_1.

The value of T_1 ranges broadly, depending on the particular type of nucleus, the nucleus' location within a molecule, and the physical state of the sample (solid or liquid). For liquids or solutions, values of $10^{-2}-10^2$ seconds are typical. For crystalline solids, T_1 values are much longer (Section 12-3). For now, just remember that the larger the value of T_1, the longer a collection of nuclei takes to reach (or return to) thermal equilibrium.

There is another reason why the magnitude of T_1 is important. Suppose we have a Boltzmann distribution of nuclei precessing in a magnetic field, and we irradiate the collection with photons of precisely the correct frequency (and energy) to cause transitions between the lower ($m = +\frac{1}{2}$) level and the upper ($m = -\frac{1}{2}$) level. Because there is initially such a small difference between the populations of the two states, it won't be long before the populations are equalized through the absorption of the photons! This, of course, means the spin system has become saturated and no further net absorption is possible. However, if we turn *off* the source of rf radiation, the system can relax back to the Boltzmann distribution (at a rate controlled by T_1) and absorption can be resumed. This fact presents us with a paradox. The spin system must absorb enough radiant power for us to be able to detect the process instrumentally, but not so much as to cause saturation.

This brings up the subject of how an nmr signal is actually generated. Figure 2-7 depicts a collection of $I = \frac{1}{2}$ nuclei within a magnetic field aligned with the $+z$ axis. Before irradiation, the

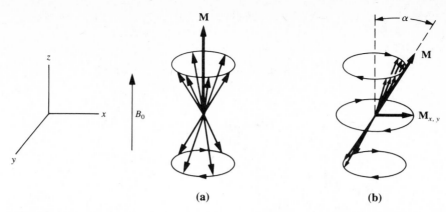

FIGURE 2-7 Precession of nuclei around an external magnetic field. Boldface arrow **M** represents net magnetization. (**a**) Before irradiation; (**b**) during irradiation.

nuclei in both spin states are precessing with characteristic frequency, but they are completely out of phase, i.e., randomly oriented around the z axis. The net nuclear magnetization **M**, the vector sum of all the nuclear magnetic moments, is aligned statically along the z axis, with no component in the x,y plane. Now, a rather strange thing happens when irradiation begins: all of the individual nuclear magnetic moments become *phase coherent*. That is, they track the oscillating magnetic field of the radiation and form a precessing "bundle" as shown in Figure 2-7b. Provided we haven't saturated the system by absorbing too many photons, this phase coherence forces the net magnetization vector **M** to precess around the z axis with the characteristic Larmor frequency. As such, **M** has a component in the x,y plane ($\mathbf{M}_{x,y}$) also oscillating with the same frequency. The tip angle α that **M** makes with the z axis controls the magnitude of $\mathbf{M}_{x,y}$, by the relation

x,y COMPONENT OF M $$M_{x,y} = M \sin \alpha \qquad\qquad (2\text{-}10)$$

The angle α is, in turn, determined by the power and duration of the electromagnetic irradiation.

Because we now have a magnetic field oscillating in the x,y plane, we can make use of Faraday induction to produce an nmr signal. By orienting a receiver coil of conducting wire with its axis in the x,y plane, the oscillating magnetic field of the nuclei induces an alternating current within the wire. And the frequency of this alternating current is exactly the precessional frequency of the nuclei. All we need do is amplify this signal and plot it as a function of irradiation frequency, and we have generated an nmr spectrum!

Before we leave the topic of relaxation, however, there is one other type of process that should be mentioned. After irradiation ceases, not only do the populations of the $m = +\frac{1}{2}$ and $m = -\frac{1}{2}$ states revert to a Boltzmann distribution, but also the individual nuclear magnetic moments begin to lose their phase coherence and return to a random arrangement around the z axis. This latter process, called **spin–spin** (or **transverse**) **relaxation**, causes decay of $\mathbf{M}_{x,y}$ at a rate controlled by the **spin–spin relaxation time** T_2. Usually T_2 is considerably shorter than T_1. A little thought should convince you that if $T_2 < T_1$, then

- Spin–spin (dephasing) relaxation takes place much faster than spin–lattice (Boltzmann distribution) relaxation.

As we shall discover in Chapter 3, the line width of nmr spectral signals depends intimately on the magnitude of T_2.

SUMMARY

1. The nucleus of an atom consists of a number (Z) of protons and a number (N) of neutrons. The atomic number Z determines the identity of the nucleus, while the sum $Z + N$ determines the nominal mass (A) of the nucleus.
2. Isotopes of a given element have the same value of Z but different values of N and A.
3. Nuclear spin (I) is a property characteristic of each isotope and is a function of Z and N. The values

of I can only be zero, n (an integer), or $n/2$ (where n is an odd integer). Only if $I \neq 0$ can the isotope be studied by nmr methods. The two most frequently studied nuclei are ^{1}H and ^{13}C.

4. Each isotope with $I \neq 0$ has a characteristic magnetogyric ratio (γ) that determines the frequency of its precession in a magnetic field of strength B_0:

$$v_{\text{precession}} = \frac{\gamma B_0}{2\pi} \qquad (2\text{-}6)$$

It is this frequency that must be matched by the incident electromagnetic radiation for absorption to occur.

5. When a collection of nuclei with $I \neq 0$ is immersed in a strong magnetic field, the nuclei distribute themselves among $2I + 1$ spin states (orientations), each with its own value of magnetic spin quantum number m_i. Nuclei in each spin state precess at the same frequency.

6. The energy of the ith nuclear spin state is given by

$$E_i = \frac{-m_i B_0 \gamma h}{2\pi} \qquad (2\text{-}3)$$

7. The relative population of each spin state is determined by the Boltzmann distribution:

$$\frac{P_{m_j}}{P_{m_i}} = \exp\left(\frac{-\Delta E_{ij}}{kT}\right) \qquad (2\text{-}8)$$

Under conditions of a typical nmr experiment, this ratio of spin state populations is near unity, differing only by a few parts per million.

8. If the two (or more) spin state populations become equal, the system is said to be saturated, and no net absorption can occur.

9. Under conditions of an nmr experiment, nuclei can readjust their nuclear spin orientations through two types of relaxation processes. Spin–lattice (longitudinal) relaxation (governed by relaxation time T_1) involves the return, after irradiation, of the nuclei to a Boltzmann distribution. Spin–spin (transverse) relaxation (governed by T_2) involves the dephasing of the "bundled" nuclear spins after irradiation is stopped.

3 OBTAINING AN NMR SPECTRUM

3-1. A Typical NMR Spectrometer

From the foregoing discussion, we already have a notion about the basic components of an nmr spectrometer. There will be a magnet, a source of rf radiation, a sensing coil, the electronics to amplify the signal, a plotter to provide hard copies of the resulting spectra, and, of course, the sample. In this chapter we will refine our spectrometer design as we consider its performance and limitations.

A. The magnet

Magnets are of three general types: permanent magnets, electromagnets, and superconducting magnets. There are advantages and disadvantages with each type. Permanent magnets are less costly, they have relatively stable fixed magnetic fields, and they require no electric current to generate the field. But the strength of their fields is limited. Electromagnets, on the other hand, are more costly to build and operate, but their field strength is variable and they can generate fields much stronger than permanent magnets can. Permanent magnets in older nmr spectrometers had field strengths of ~ 1.4 T, while today's superconducting electromagnets reach field strengths of up to 14 T; for comparison, the Earth's permanent magnetic field averages 0.00006 T.

Because the precessional (and resonance) frequency of a nucleus is directly proportional to the magnetic field strength [Eq. (2-6)], so is the *difference* Δv between resonance frequencies of nonidentical nuclei:

FREQUENCY DIFFERENCE $$\Delta v = v_1 - v_2 = \frac{\gamma_1 B_0}{2\pi} - \frac{\gamma_2 B_0}{2\pi} = \frac{(\gamma_1 - \gamma_2)B_0}{2\pi} \qquad (3\text{-}1)$$

It is therefore advantageous to use the strongest possible magnetic field to obtain the greatest separation (and hence, resolution) between signals. Remember also that a stronger field results in larger energy gaps between spin states [Eq. (2-4)] and hence larger populations in the lower state [Eq. (2-8)]. This fact helps enhance the intensity of an nmr signal, which, it turns out, is approximately proportional to the square of B_0.

The two most important characteristics of the magnet in any nmr spectrometer are the stability and homogeneity of its magnetic field. The field produced by a magnet is highly sensitive to the magnet's temperature. Permanent magnets are cooled by ambient air, and since there is no electrical (resistive) heating, satisfactory stability can be achieved by simply controlling the temperature of the surroundings. A conventional electromagnet, however, produces a significant amount of heat as a direct result of the current that produces its field. The temperature of such magnets is controlled by cooling coils that permeate the magnet itself, through which thermostated chilled water is circulated. With superconducting magnets, liquid helium maintains the temperature of the special conducting coils at 4 K, where the electrical resistance (and hence resistive heating) of the conductors is zero.

In the case of electromagnets there must also be some mechanism by which the strength of the magnetic field is controlled. This can be achieved through an electronic technique known as **locking**. To accomplish this, we must first select a substance with the capacity to give a strong nmr signal separate from those of the sample. If this substance is kept physically apart from (but close to!) the sample, it is referred to as an **external lock**. More commonly it is dissolved within the sample itself and termed an **internal lock**. In either case, the frequency of the lock signal is electronically compared to a constant frequency from an rf oscillator, and, by means of a feedback loop to the circuit generating the magnetic field, the field is automatically and continuously adjusted to keep the frequencies equal.

EXAMPLE 3-1 Suppose the lock signal frequency is found to be less than the constant frequency of the rf oscillator. Should the magnetic field be increased or decreased to bring it back to the nominal value?

Solution: Remember [Eq. (3-1)] that the frequency of any signal increases in direct proportion to the field strength. Thus, to *increase* the lock signal frequency, we need to *increase* the field strength.

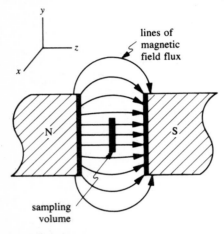

FIGURE 3-1 Nonhomogeneity of the field between the pole faces of a magnet (exaggerated).

Once a stable field is established, the question remains as to whether that field is completely *homogeneous* (uniform) throughout the region between the pole faces of the magnet. It is clear from Figure 3-1 that, in fact, the field is not uniform, except near the very center of the pole gap, and it is here we will place the sample. The degree of homogeneity required for a given nmr experiment depends on the desired level of resolution, which, in turn, controls the precision of the measurement. In the case of ^1H nuclei at 2.35 T, for example, Eq. (2-6) shows that to achieve a precision of ± 1 Hz at a frequency of 100 MHz (a precision level of ten parts per billion!), the field must be homogeneous to the extent of $\pm 2.35 \times 10^{-8}$ T. Such phenomenal uniformity, even at the center of the field, can be achieved only by means of two additional techniques. First, the sample vessel (normally a precisely constructed glass tube) is oriented along the y axis in Figure 3-1 (this axis is relabeled the z axis in the case of superconducting magnets) and is spun around its axis by means of an air stream turning a paddle wheel attached to the tube. This spinning helps to "average out" any slight inhomogeneities of the field in the sample region. Second, the contour of the field itself can be varied (within *very* narrow limits) by passing extremely small currents through **shim coils** wound around the magnet itself. The small field gradients caused by currents flowing through these shim coils can be adjusted to improved the uniformity of the field further. This process, called shimming or tuning the magnet, was accomplished on older instruments by manual adjustment of a series of shim controls (labeled x, y, z, xy, yz, xz, curvature, etc.) before each spectrum was recorded. With the advent of microprocessor-controlled spectrometers, the bulk of these adjustments can now be accomplished automatically.

EXAMPLE 3-2 If the Earth's magnetic field varied by ± 0.00001 T, what level of error would this introduce in the resonance frequency of ^1H nuclei at 2.35 T?

Solution: We can use a simple proportion:

$$\frac{0.00001 \text{ T}}{2.35 \text{ T}} = \frac{\Delta v}{100 \text{ MHz}}$$

$$\Delta v = 0.000430 \text{ MHz} = 430 \text{ Hz}$$

As we will later see, this would be a *very* large error!

We are left with a paradox. To get an acceptable signal, we need to have as many nuclei as possible in the sample region (remember the minute difference between spin state populations?). But now we find we must restrict our sample to the smallest possible volume in the interest of precision and resolution. The eventual compromise results in a sample cavity of ~ 0.05 cm³ for instruments that accommodate narrow-bore tubes to ~ 1.0 cm³ for instruments using wide-bore tubes. [However, for medical use, some low-resolution nmr instruments have been constructed that accommodate entire human bodies! (see Section 12-4)].

EXAMPLE 3-3 A typical narrow-bore nmr tube has an outside diameter of 5.00 mm (0.500 cm) and an inside diameter of 4.20 mm. It is usually filled to a height of ~ 2.5 cm. (a) What is the wall thickness of the glass tube? (b) If the sample cavity (including the glass of the tube) is 0.050 cm³, what are the dimensions (radius, height, and volume) of the volume that is actually occupied by sample? [*Hint:* Volume of cylinder = $\pi r^2 h$.]

Solution: Refer to Figure 3-2.

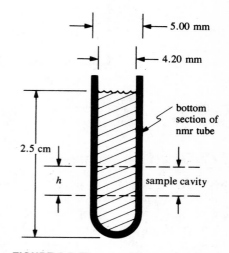

(a) wall thickness $= \dfrac{5.00 \text{ mm} - 4.20 \text{ mm}}{2} = 0.40$ mm

(b) The diameter of the sample is the same as the inside diameter of the tube, 0.42 cm. The height of the sample cavity is given by

$$h = \frac{V_{\text{total}}}{\pi r_{\text{total}}^2} = \frac{0.050 \text{ cm}^3}{\pi (0.25 \text{ cm})^2} = 0.255 \text{ cm}$$

Since $r_{\text{sample}} = 0.42$ cm$/2 = 0.21$ cm,

$$V_{\text{sample}} = \pi r_{\text{sample}}^2 h = \pi (0.21 \text{ cm})^2 (0.255 \text{ cm})$$
$$= 0.035 \text{ cm}^3$$

Thus, only 70% of the sample cavity is filled with sample in this case. Clearly, the thinner the wall thickness of the tube, the more sample actually gets into the sample cavity.

FIGURE 3-2 The sampling volume of an nmr spectrometer.

B. Assembling the pieces

Our perseverance is about to be rewarded! Figure 3-3a depicts the basic features of a typical nmr spectrometer. Notice the components we've previously discussed: the magnet (with its field defining the z axis) and lock system, a sensing (or receiver) coil around the y axis, an amplifier and plotter, the sample (including the lock substance), and a transmitter coil (around the x axis), which serves as the source of rf radiation. The two coils and the sample cavity are referred to collectively as the **probe**.

A brief review of Chapters 1 and 2 should convince you of the necessity for this particular arrangement of the transmitter and receiver coils. The nuclei will precess around the z axis (horizontal in Figure 3-3a, but vertical in the figures in Chapter 2). In order to cause resonance, the magnetic vector of the rf radiation must oscillate in the x,y plane, and this is accomplished by passing an rf alternating current through a coil whose axis (here, the x axis; see Figure 3-3b) is perpendicular to the z axis. Then, as we saw in Section 2-3B, the resulting precession of x,y magnetization around the z axis (Figure 2-7) induces a Faraday current in the receiver coil, whose axis also lies in the x,y plane. To minimize direct electronic interference between the two circuits, the transmitter and receiver coils are perpendicular to each other.

We're now ready. The magnetic field is stabilized at 2.35 T and a sample containing ¹H nuclei is spinning in the probe. We activate the rf oscillator, producing continuous 100.0 MHz radiation, and… yes! A weak signal is detected on the recorder! Exciting experiments just like this in the separate laboratories of Block and Purcell marked the true discovery of nmr. For this, they shared the 1952 Nobel Prize in physics. But let's not celebrate yet. What has this told us, except that we have ¹H in the sample? Certainly, we'd like to be able to adjust parameters to test for other nuclei besides ¹H. How can we accomplish that?

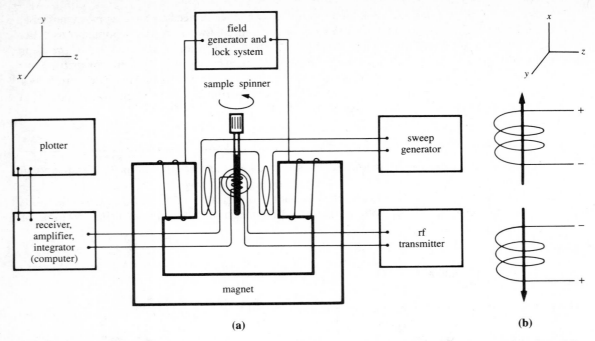

FIGURE 3-3 (**a**) Basic design of an nmr spectrometer. (**b**) The rf magnetic field direction as a function of the direction of rf current in the transmitter coil.

3-2. The Continuous-Wave Experiment

Recall from Chapter 2 that, at a given field strength, each different nucleus (with $I \neq 0$) precesses at a unique frequency governed by its magnetogyric ratio [Eq. (2-6)]. Thus, if we intend to use our newly created spectrometer to observe nuclei other than simply ^{1}H, there must be a way to vary (or *sweep*) either the operating rf frequency or the magnetic field strength. In fact, both techniques are possible. But before we describe them, remember that we are supplying the rf radiation continuously (nonstop). Therefore, these are referred to as continuous-wave (cw) techniques.

A. The frequency-sweep mode

In the case of a frequency-sweep cw experiment, the magnetic field strength (B_0) is kept constant. If our sample contains two different types of nuclei (A and B) characterized by magnetogyric ratios γ_A and γ_B ($\gamma_A > \gamma_B$), the energies of the nuclei in their two spin states is as depicted in Figure 3-4a.

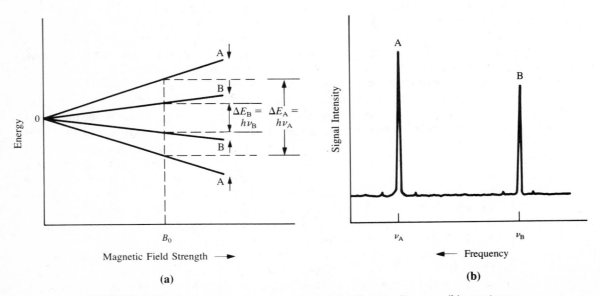

FIGURE 3-4 The frequency-sweep experiment. (**a**) Energy diagram; (**b**) spectrum.

In order for A and B to generate resonance signals, Eq. (2-7) requires that the nuclei be irradiated with frequencies given by

RESONANCE FREQUENCIES

$$\nu_A = \frac{\Delta E_A}{h} \tag{3-2a}$$

$$\nu_B = \frac{\Delta E_B}{h} \tag{3-2b}$$

where $\nu_A > \nu_B$. Thus, to generate spectral signals from both nuclei, it is necessary to sweep the operating frequency over the range ν_A to ν_B (or vice versa). This range of frequencies is called the **spectral** (or sweep) **width** (SW). So, we will make a small addition to the design in Figure 3-3a: a continuously variable rf transmitter. Beginning at a value *less* than ν_B, we will gradually increase the frequency of the continuous rf radiation to a value above ν_A while constantly monitoring the receiver circuit for a signal. Lo and behold, as the frequency passes through ν_B, then ν_A, two beautiful signals appear on the chart paper. Figure 3-4b is our first bona fide nmr spectrum, with frequency increasing from *right to left* (as is the convention). Also, as we will see in Section 5-1A, the intensity of each signal is proportional to the number of nuclei giving rise to the signal.

B. The field-sweep mode

Although some nmr spectrometers do employ the above frequency-sweep (at constant field) technique, it turns out to be technically simpler to maintain a constant operating rf frequency and vary the magnetic field to achieve resonance for each type of nucleus in the sample. By passing an appropriate current though the **sweep coils** in Figure 3-3a, the magnetic field can be varied in a well-defined way, while maintaining its homogeneity in the sample area. Let's investigate how this change affects the spectrum.

Because the frequency (ν_{photon}) of rf radiation is now constant, only photons of energy $h\nu_{photon}$ are available in the experiment [see Eq. (1-3)]. Thus, we must now adjust the *energy gap* [ΔE, Eq. (2-4)] for each nucleus by varying the magnetic field strength B. By rearranging Eq. (2-7) we can solve for the field strength at which nuclei A and B will enter resonance:

FIELD STRENGTH AT RESONANCE

$$B_A = \frac{2\pi\nu_{photon}}{\gamma_A} \tag{3-3a}$$

$$B_B = \frac{2\pi\nu_{photon}}{\gamma_B} \tag{3-3b}$$

Figure 3-5a depicts the situation graphically. This time, instead of varying the frequency (photon energy) until it matches the nuclear energy gap, we vary the energy gap (by varying the field strength) until it matches the photon energy. But most importantly, the resulting spectrum (Figure 3-5b) is

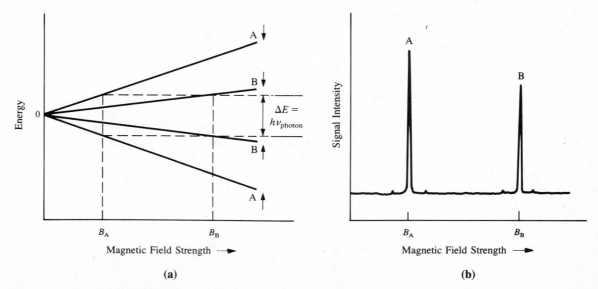

FIGURE 3-5 The field-sweep experiment. (a) Energy diagram; (b) spectrum.

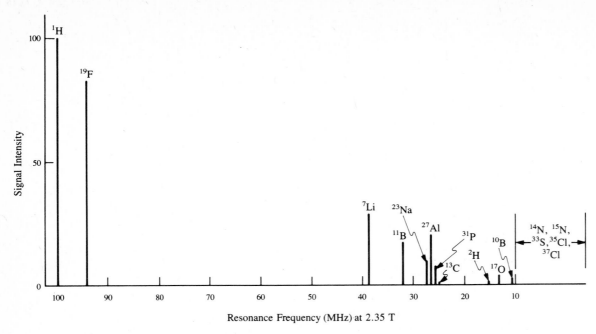

FIGURE 3-6 A heteronuclear nmr spectrum.

indistinguishable from the frequency sweep spectrum, except that the abscissa is now calibrated in magnetic field units increasing from *left to right*.

Comparing Figure 3-4b with Figure 3-5b reveals that nucleus A enters resonance at higher frequency (constant field) or *lower* field (constant frequency) than does nucleus B. For this reason, the right-hand (low-frequency) end of an nmr spectrum is called the *high-field* (or *upfield*) end; the left-hand (high-frequency) end is called the *low-field* (or *downfield*) end.

Let's now prepare a sample with equal numbers of each nucleus represented in Table 2-1 and record the sample's nmr spectrum at 2.35 T. The result would be as shown in Figure 3-6. Of course, we could substitute units of magnetic field strength on the abscissa if we wished, but the convention is to calibrate it in units of frequency, regardless of whether the experiment involved field sweep or frequency sweep. But remember: Both frequency-sweep and field-sweep experiments require continuous application of rf radiation (of either fixed or varying frequency), and therefore both are continuous-wave techniques.

EXAMPLE 3-4 Suppose that, at a nominal field strength of 2.35 T, nmr signals were detected at 100.0 and 25.2 MHz. **(a)** What are the two nuclei giving rise to the signals? **(b)** If the spectrum were obtained by the field-sweep method, what range of field strength would be required at an operating frequency of 100.0 MHz?

Solution:

(a) From Table 2-1 and Example 2-6, we know the nuclei are ^{1}H and ^{13}C, respectively.

(b) At an operating *frequency* of 100 MHz, ^{1}H nuclei enter resonance at a field strength of 2.35 T. But at what field strength will ^{13}C nuclei be found? Use Eq. (3-3) and the value of γ for ^{13}C from Table 2-1.

$$B_{^{13}C} = \frac{2\pi\nu_{\text{photon}}}{\gamma_{^{13}C}}$$

$$= \frac{2(3.14 \text{ rad})(100 \times 10^6 \text{ s}^{-1})}{67.264 \times 10^6 \text{ rad T}^{-1} \text{ s}^{-1}}$$

$$= 9.33 \text{ T}$$

Thus, to observe *both* ^{1}H and ^{13}C, we would need to be able to vary the field strength from 2.35 T to 9.33 T, or by nearly 7 T. For reasons described a little later, this is impossible to accomplish with any measure of precision.

C. Spinning side bands and the signal-to-noise ratio

A close examination of the spectral peaks in Figures 3-4b and 3-5b brings up some additional considerations. Figure 3-7 is an enlargement of one of the peaks. You may have noticed the two small signals symmetrically flanking each large signal in Figures 3-4b and 3-5b. These so-called **sidebands** or **satellite peaks**, marked with asterisks in Figure 3-7, are an unavoidable result of spinning the sample to improve field homogeneity (see Section 3-1A). If the field has been well shimmed, such **spinning sidebands** rarely exceed 1% of the height of the main peak. Moreover, they will always be separated from the main peak by *exactly* the spin rate in Hz. This is, in fact, the most direct way to measure the sample spin rate. One can, of course, completely eliminate spinning side bands by turning off the spinner. But this would cause substantial broadening of the signal (with an accompanying decrease in signal height) because of the small magnetic field inhomogeneities.

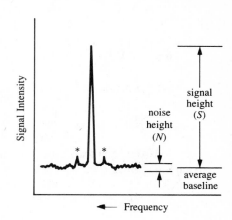

FIGURE 3-7 An nmr signal showing sidebands (*) and signal-to-noise ratio.

EXAMPLE 3-5 The two side bands in Figure 3-7 are each separated from the main peak by 58 Hz. **(a)** What is the spin rate of the sample tube? **(b)** How could you confirm that these peaks were indeed spinning side bands?

Solution:

(a) Because the frequency separation of a sideband from the main peak is equal to the spin rate, the latter must be 58 Hz.
(b) Varying the spin rate (by adjusting an air flow directed at the spinner paddle) causes spinning side bands to shift position accordingly. If the peaks do *not* shift position, they are *not* spinning side bands. There *are* other types of satellite peaks that we will encounter later.

You will also notice in Figure 3-7 that the baseline is not perfectly flat. Instead, there is perceptible random background noise causing a continuous wiggle in the baseline. Even in the most carefully built and tuned spectrometers there will always be some electronic noise generated by the various circuits within the instrument. Normally, this does not present a problem, provided the sample gives signals strong enough to be easily differentiated from the noise. But what about really weak signals? For example, a sample of natural hydrogen (Table 2-1) contains 99.985% ^{1}H with a relative sensitivity of 1.00, but a sample of natural carbon contains only 1.1% ^{13}C, which has a relative sensitivity of 1.59×10^{-2}. These two factors—low natural abundance and low relative sensitivity—combine to make the carbon signal only about 2×10^{-4} times as intense as a hydrogen signal for equal numbers of atoms of each element. Thus, a ^{13}C signal is much harder to distinguish from noise than an ^{1}H signal.

To circumvent this problem, let's define a quantity called **signal-to-noise (S/N) ratio** where S and N are approximated as their respective peak heights, as shown in Figure 3-7.

EXAMPLE 3-6 Using a ruler, estimate the S/N ratio in Figure 3-7.

Solution: The exact values of S and N will depend on the scale of your ruler. But the ratio of the two numbers, S/N, should come out to about 19.8.

The goal in an nmr experiment is to maximize the S/N ratio, and this goal can be achieved in several ways. One method is to scan (vary the field or frequency) very slowly and use a series of rf filters to remove some of the noise. While this technique results in some improvement, care must be taken to use low rf power levels to avoid saturation problems (see Section 2-3A). A far better improvement can be obtained by relying on a result from information theory. This theory tells us that, if the spectrum is scanned repeatedly and the resulting spectral data are added together, signal intensity will increase directly with n, while noise (being random) will only increase by $\sqrt{n}$, where n is the number of scans.

SIGNAL-TO-NOISE RATIO $$S/N = (S/N)_1\left(\frac{n}{\sqrt{n}}\right) = (S/N)_1\sqrt{n} \qquad (3\text{-}4)$$

where $(S/N)_1$ is the ratio after a single scan. Thus, the S/N ratio improves linearly with $\sqrt{n}$.

To make use of this multiple scan technique, we need a computer to collect the spectral data (digitally) from each scan, add the data together, then divide the sum by n. The technique is known as signal averaging, or CAT, for computer-averaged transients. Because each scan may require ~ 10 minutes, this process can take quite a long time, during which the magnetic field must be kept as perfectly stable and homogeneous as possible.

EXAMPLE 3-7 A single nmr scan of a certain highly dilute sample exhibits a S/N ratio of 1.9. If each scan requires 10 minutes, what is the minimum time required to generate a spectrum with a S/N ratio of 19?

Solution: We can rearrange Eq. (3-4) to calculate the required number of scans:

$$\sqrt{n} = \frac{S/N}{(S/N)_1} = \frac{19}{1.9} = 10$$

$$n = (10)^2 = 100$$

The time required for 100 scans is

$$t = (100 \text{ scans})(10 \text{ min per scan}) = 1000 \text{ min} = 16 \text{ hr } 40 \text{ min!}$$

3-3. The Pulsed Fourier Transform Technique

Further advances in S/N ratio improvement had to await the development of faster computer microprocessors, which was exactly what happened during the 1970s. Armed with very fast and efficient microcomputers with large memories, chemists could adapt the **pulsed Fourier transform** (PFT) technique to nmr. Before proceeding with this section, you may want to review Sections 2-2B and 2-3A and B.

A PFT experiment (which is performed at constant magnetic field), differs from a continuous-wave experiment in that the rf radiation is supplied by a brief but powerful pulse of rf current through the transmitter coil. This pulse, centered at frequency ν_0, is so intense and covers such a large span of frequencies that it allows absorption by all the nuclei within the **spectral width** (SW) simultaneously (see Figure 3-8). The duration of the pulse (t_p) determines the frequency range covered. SW is equal to t_p^{-1}, a direct consequence of Eq. (1-5). The reason for setting ν_0 at the edge (rather than the center) of the spectral region of interest is that by doing so we collect data that depend on the difference actual precession frequency and ν_0 ($\nu_{\text{precession}} - \nu_0$).

Let's first consider the consequences of this pulse on a collection of identical nuclei. As a result of the magnetic field in which they are immersed, the nuclei are initially precessing with characteristic frequency. When they receive the pulse,

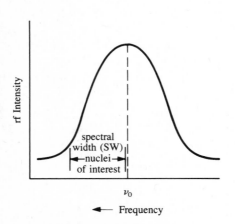

FIGURE 3-8 Pulse profile.

they form a phase-coherent bundle (Section 2-3B), with a component of magnetization precessing in the x,y plane. As in a continuous-wave experiment, this induces an rf current in the receiver coil with the same characteristic frequency. The intensity of this induced current is directly proportional to the magnitude of the x,y magnetization, which is, in turn, controlled by the tip angle α [Figure 2-7 and Eq. (2-10)].

As stated in Chapter 2, the magnitude of α is determined by the power and duration of the irradiating pulse, as well as by the frequency difference $(\nu_{\text{precession}} - \nu_0)$. In most cases the optimum signal is obtained from a 90° tip angle, and a pulse of the length required to produce this value of α is called a **90° pulse**. The pulse duration (t_p) required to accomplish this 90° flip can be estimated from Eq. (3-5),

PULSE DURATION $$t_p \leq (4\,\text{SW})^{-1} \tag{3-5}$$

but usually a process of trial and error is required to provide the best S/N ratio. Pulse durations on the order of 10 μs are typical.

The next step in the PFT process is to monitor the induced ac receiver signal (voltage or current) as a function of *time*. In order to enable us to carry out multiple pulse averaging, the data must be collected digitally. Thus, the microprocessor samples the voltage in the receiver coil at regular intervals (called **dwell time**, t_d). A typical set of such data might look like Figure 3-9. The pattern described by these points is not immediately obvious because data points have not been collected often enough, at least for our eyes to recognize it. The experiment can be repeated with the dwell time reduced by a factor of five, thus affording five times as many data points in the same total time. Figure 3-10 shows the result: a more

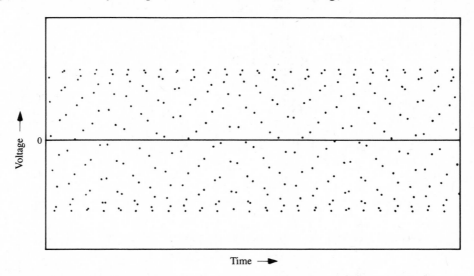

FIGURE 3-9 Voltage induced in the receiver coil, collected digitally as a function of time.

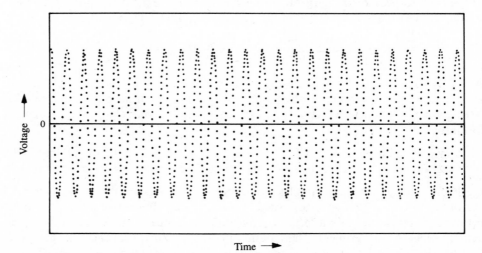

FIGURE 3-10 Effect of dwell time on the signal. Compare with Figure 3-9.

detailed and easily recognizable cosine wave. So, one important lesson is that the voltage must be sampled often enough to make the pattern recognizable. It turns out that to find the frequency of a wave by this method, we must acquire at least two points per cycle. Therefore, so that frequencies as high as the spectral width can be determined, dwell time should be set to the inverse of twice the spectral width.

DWELL TIME $$t_d \geq (2\,SW)^{-1} \tag{3-6}$$

Notice also that the inverse of the dwell time is the number of data points collected per second.

Actually, the data would more closely resemble the pattern in Figure 3-11. Such a pattern is called a **modulated free induction decay (FID) signal** (or **time-domain spectrum**) because the current intensity decreases with time. This decay is the result of spin–spin (or transverse) relaxation (Section 2-3B), which reduces the net magnetization in the x,y plane. The envelope of the damped cosine wave in Figure 3-11 describes an exponential decay function whose decay time is T_2^*, the *effective* spin–spin relaxation time. T_2^* differs from T_2 (Section 2-3B) in that it includes transverse relaxation due to field inhomogeneity (usually the dominant contributor), as well as inherent spin–spin relaxation. As a result, T_2^* is always shorter (implying faster relaxation) than T_2.

Still, the important information is the as-yet-unknown precessional frequency of the nuclei. How is that information obtained from the data in Figure 3-11? First, note that the *frequency* of the cosine wave

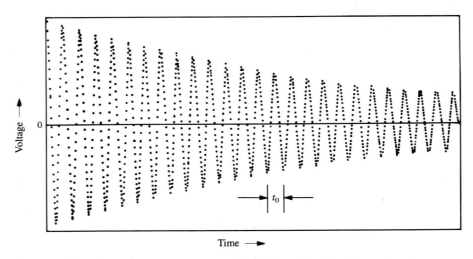

FIGURE 3-11 The complete modulated free induction decay signal.

is unaffected by the exponential decay (compare Figures 3-10 and 3-11). Therefore, all that is needed is to measure the peak-to-peak or trough-to-trough time (t_0 in Figure 3-11). Suppose this length of time were 2.0×10^{-2} s. Then the frequency of the cosine wave [see Eq. (1-1)] would be

$$\nu = \frac{1 \text{ cycle}}{2.0 \times 10^{-2}\,s} = 5.0 \times 10^1 \text{ Hz} = 50 \text{ Hz}$$

This frequency, extracted from the FID data, is actually a frequency *difference*, i.e., $\nu_{precession} - \nu_0$. In a sense, we have just performed a time-to-frequency transformation on the data: we have taken digital data describing a periodic function of time and extracted the function's characteristic frequency. And most importantly,

- The transformed signal generated in a PFT experiment is identical to the signal observed in a CW experiment (e.g. Figure 3-7) and can be generated in a PFT experiment from a single rf pulse.

In experiments with either dilute samples or insensitive nuclei (such as ^{13}C), one pulse usually does not give a S/N ratio high enough to allow us to determine the signal's frequency accurately. The S/N ratio can be improved by repeating the pulse/data acquisition sequence then adding the new data to the original data, as in a CAT experiment. The required number of pulse sequences is determined by the desired S/N ratio. But, there is an additional consideration: To avoid saturation one must allow enough time between pulses for the nuclei to reach (or nearly reach) their original equilibrium (Boltzmann) distribution. Therefore, this **delay time** (t_w) is a function of T_1 (Section 2-3B).

Most of the molecules examined by nmr have several sets of nuclei, each with a different precessional frequency. Furthermore, each set has different relaxation times (T_1 and T_2), and usually there are different numbers of nuclei within each set. These factors combine to give a very complex digital FID curve consisting of N data points (numbered from 0 to $N - 1$). For example, Figure 3-12 shows the FID curve for the ^{1}H nuclei of toluene (structure **3-1**). At this point it becomes necessary for the computer to

3-1

recognize the patterns mathematically and extract the signal frequency and relative signal intensity for each set of nuclei. The computer accomplishes this analysis by performing a **Fourier transformation** of the FID data according to

FOURIER TRANSFORMATION
$$F_j = \sum_{k=0}^{N-1} T_k e^{(-2\pi i jk/N)} \qquad \text{for } j = 0 \text{ to } N - 1 \tag{3-7}$$

where T_k represents the kth point of the time-domain (FID) data, and F_j represents the jth point in the resulting frequency-domain spectrum. Thus, *each* F_j point (from 0 to $N - 1$) requires a summation over all N points in the FID curve, and the total calculation involves N^2 multiplications and additions. An algorithm developed by Cooley and Tukey simplified this extremely time-consuming calculation, bringing it within the capability of modern microprocessors. Today, the transformation takes only a few seconds, after which the frequency domain spectrum F_j can be plotted.

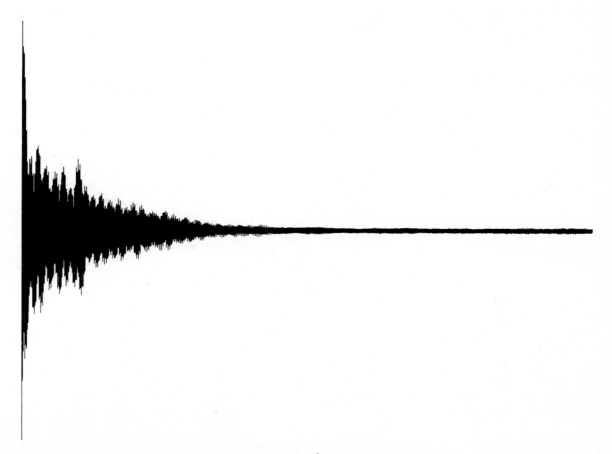

FIGURE 3-12 The 300-MHz ^{1}H FID curve for toluene.

As a typical example of the pulse/data acquisition/delay/repeat sequence, let's consider the parameters we must specify when we generate a PFT ^{1}H nmr spectrum. First, we select a spectral width (SW, Figure 3-8) that covers the nuclei of interest, typically 1000 Hz in the case of ^{1}H at 100 MHz. Remember that SW determines the pulse duration [Eq. (3-5) and Section 3-4A]. Since the frequency of each component wave can be uniquely determined by two points per cycle, and our ^{1}H spectral width might include frequencies $(v - v_0)$ in the range 0–1000 Hz, we must sample points twice that fast, or 2000 Hz (a dwell time of 500 μs), to ensure generating all relevant spectral frequencies [refer to Eq. (3-6)]. Moreover, we are limited by our computer's memory capacity. Generating 2000 points per second will fill up 16K (16,384 locations) of computer memory in 8 seconds. This fact determines the **acquisition time** (t_{acq}), the length of time a given FID signal is actually monitored. Resolution (r), the ability to distinguish two nearby signals, is inversely proportional to acquisition time. Thus, the 8-second acquisition

SPECTRAL RESOLUTION $\qquad\qquad\qquad r = (t_{acq})^{-1}$ $\qquad\qquad\qquad\qquad\qquad$ **(3-8)**

time would provide a resolution of $(8\ \text{s})^{-1} = 0.125$ Hz. Note that the smaller the value of r, the *better* the resolution. But because the signal decays as the nuclei lose their coherence, there is a time limit beyond which further monitoring of the FID provides more noise than signal. The usual compromise is to have a short enough *dwell time* to cover the spectral width, and a long enough *acquisition time* to provide the desired resolution, consistent with the computer's memory limits.

After collecting data from one pulse, we must wait enough time after the pulse for the nuclei to relax to equilibrium. A total of at least $3T_1$ is usually adequate, but part of this is spent as acquisition time. Thus, the *delay time* (t_w) can be calculated as

DELAY TIME $\qquad\qquad\qquad\qquad t_w = 3T_1 - t_{acq}$ $\qquad\qquad\qquad\qquad\qquad$ **(3-9)**

Since ^{1}H nuclei normally exhibit T_1 values on the order of 1 second, to generate an ^{1}H spectrum we need essentially no additional delay after the acquisition time. For ^{13}C and other nuclei, however, long delay times are sometimes required. Following the delay, the sample is irradiated with another pulse, and the data acquisition sequence begins anew. Figure 3-13 depicts this entire process. After some number of these pulse sequences (depending on sample concentration and type of nuclei), the data are subjected to a Fourier transformation, and the S/N ratio of the resulting frequency-domain spectrum is ascertained. If it is not adequate, the pulse sequence is resumed until the desired S/N ratio is attained.

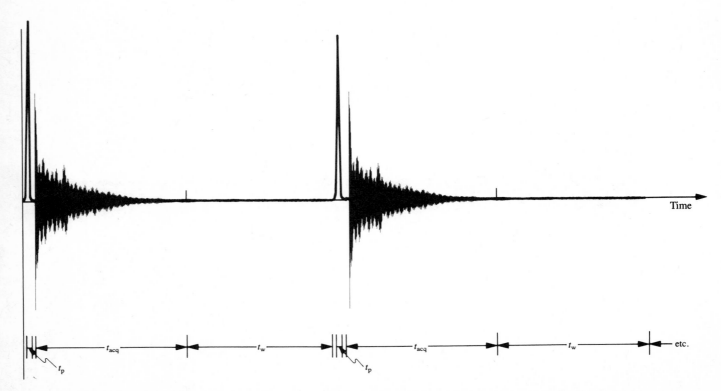

FIGURE 3-13 The PFT data acquisition sequence, showing pulse duration (t_p), acquisition time (t_{acq}), and delay time (t_w).

A PFT experiment thus encompasses two procedures:

(**1**) The collection of data by a pulse/data acquisition/delay sequence, repeated enough times to yield an FID signal possessing the desired S/N ratio, and

(**2**) Fourier transformation of the FID data, followed by plotting of signal intensity as a function of frequency.

Still, in principle

- The frequency-domain spectrum generated by the PFT technique is essentially indistinguishable from that obtained by continuous-wave methods.

EXAMPLE 3-8 Suppose you are to examine a ^{13}C spectral window of 5000 Hz. The nuclei have T_1 values of ~ 2 s and your computer memory has a capacity of 32K (32,768 locations). The desired resolution is 0.5 Hz. (**a**) Using Figure 3-13 as a guide, suggest values for the various data acquisition parameters. (**b**) How many data points will you collect from one pulse sequence with these values? (**c**) If 1000 pulse sequences are required for the desired S/N ratio, how long will the total experiment required?

Solution:

(**a**) Use Eqs. (3-5), (3-6), (3-8), and (3-9).

$$t_p \leq \frac{1}{4(SW)} = \frac{1}{4(5000 \text{ Hz})} = 50 \text{ } \mu s$$

$$t_d = \frac{1}{2(SW)} = \frac{1}{2(5000 \text{ Hz})} = 100 \text{ } \mu s \ (= 10^4 \text{ data points per second})$$

$$t_{acq} = \frac{1}{r} = \frac{1}{0.5 \text{ Hz}} = 2 \text{ s}$$

$$t_w = 3(T_1) - t_{acq} = 3(2 \text{ s}) - 2 \text{ s} = 4 \text{ s}$$

(**b**) Since we will acquire 10^4 data points per second for 2 s, we'll collect 20,000 points for each pulse.

(**c**) Since the pulse time is negligible, the total pulse sequence time is essentially equal to $t_{acq} + t_w \ (= 3T_1)$; thus, each pulse sequence requires 6 seconds. If we need 1000 pulses, the total time required for the experiment is 6000 s (1 hr 40 min). Although this may seem like a long time, remember from Example 3-7 that just 100 cw scans required nearly 17 hours!

3-4. Additional Considerations

A. Line width and ringing

In Chapter 1 we found that the Uncertainty Principle establishes spectroscopic time scales. It also has a significant impact on the shape of signals. The sharpness of an nmr signal is measured by its **half-width** $v_{1/2}$, the signal's width (in Hz) at half its height (Figure 3-14). We can rewrite Eq. (1-5) in two other forms:

UNCERTAINTY PRINCIPLE

$$\Delta E \, \Delta t = h \, \Delta v \, \Delta t \geq h \qquad \textbf{(3-10a)}$$

$$\Delta v \, \Delta t \geq 1 \qquad \textbf{(3-10b)}$$

In other words, the relation says that the uncertainty in frequency is inversely related to the uncertainty in time scale. We can use the spectral half-width as a measure of uncertainty in frequency, and the nuclear relaxation time (T) as a measure of uncertainty in the lifetime of the nuclear spin states giving rise to the particular signal. Making these substitutions in Eq. (3-10b) allows us

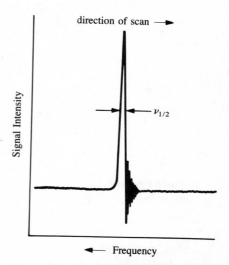

FIGURE 3-14 A typical continuous-wave nmr signal, showing half-width $(v_{1/2})$ and ringing.

to express the half-width of a spectral line as

$$v_{1/2} \geq \frac{1}{T}$$

(3-11)

Thus,

- The (half) width of a spectral signal is inversely proportional to the relaxation time of the nuclei giving rise to the signal.

Nuclei that are slow to relax (large values of T) give sharp signals; nuclei that relax rapidly give broad signals. The half-width is, therefore, controlled by the *fastest* type of relaxation (smallest value of T). For liquids and solutions, this corresponds to T_2^* values (Section 3-3), which are mainly a function of local inhomogeneities of the magnetic field. Typical half-widths so determined range from about 0.1 to 10 Hz. In the case of solid or crystalline samples there is another factor that greatly increases line widths: the direct interaction between the magnetic dipoles of neighboring nuclei. Until recently this factor, which is essentially averaged out in liquid and solution samples, precluded direct nmr investigation of solids because it broadened lines in their spectra so much. Now, however, techniques have been developed to circumvent this problem, and it is possible to obtain nmr spectra of solids routinely. This technique will be discussed briefly in Section 12-3.

The other phenomenon shown in Figure 3-14, termed **ringing**, is the symmetrically decaying wiggle pattern appearing just *after* a signal *in a continuous-wave spectrum*. The ringing is actually a beat pattern caused by interference between the fixed-frequency rf oscillator and the increasing (or decreasing) nuclear precessional frequency (a result of sweeping the field strength through the signal). It should not surprise you that the envelope of the ringing pattern describes an exponential decay controlled by T_2^*, just as with the FID signal (Section 3-3). Ringing patterns are reduced by scanning through the signal more slowly than T_2^*. They are absent in the frequency domain spectrum generated from a PFT experiment.

B. Homonuclear nmr

This chapter may have given you the impression that every nmr spectrometer is capable of scanning the entire nmr spectral region for all different nuclei. Actually, this is not the case. Because of technical limitations, it is possible to build neither a single magnet nor a single rf oscillator capable of covering the entire region in one spectrum. In fact, no nmr spectrometer is capable of examining more than one type of nucleus *at a time*! That's incredible, you say! Why buy a horribly expensive nmr spectrometer, just to find out whether you have, for example, ^1H or ^{13}C in your sample? Certainly there are far less expensive ways to get the same information. Here is the answer and this is the true reason behind the immense value of nmr:

- Although all nuclei of a given isotope (e.g., ^1H or ^{13}C, etc.) have exactly the same value of γ, there can be small variations in their exact precessional frequencies because of differences in their molecular environments.

It is these small differences that we will exploit in chapters 5–7 to gain the valuable structural information available only through nmr. Although most modern nmr spectrometers are capable of being configured for several or many different nuclei by changing probes, oscillator circuits, etc., the fact remains that

- With nmr spectroscopy we can examine only one type of nucleus at a time.

The resulting spectrum is termed a **homonuclear spectrum** because it portrays signals from only *one* type of nucleus (e.g., ^1H or ^{13}C).

This brings up the topic of cost. Commercially available nmr spectrometers are *not* inexpensive. Even the earliest 60-MHz ^1H-only instruments were priced well in excess of $20,000. Modern computer-controlled multi-nuclear PFT instruments range from about $100,000 to $500,000. A rough rule of thumb is that each MHz of operating frequency costs $1000!

C. Sampling considerations

As we saw in Section 3-2, the intensity of an nmr signal under single-scan continuous-wave conditions is determined in part by the number of nuclei within the sample volume. Because we use liquid samples or solutions, in the interest of reasonable relaxation times and half-widths, it behooves us to use the most concentrated samples possible. Example 3-3 indicated that the true sample volume in a 5-mm probe was 0.035 cm^3, which is comparable to 30 mg of typical organic compounds. Actually, however, more of the tube is filled than just the sample region, requiring total solution volumes of ~ 0.3 cm^3. In the case of solutions, we must select a solvent. There are several considerations. First, and most obviously, the solvent must dissolve the sample! Of course, if your sample is only sparingly soluble, or if there is only a fraction of a milligram of it available, you can always resort to CAT or PFT techniques to obtain the desired S/N ratio. But besides that, it is best if the solvent does *not* contain any of the particular nuclei to be examined in the experiment. For example, to obtain an ^{1}H spectrum one should pick a solvent that is free of ^{1}H. Several examples are given in Table 3-1. But suppose your ^{1}H-containing sample is soluble *only* in water? Not to worry! Simply substitute the deuterated (^{2}H, or D) analog of water (D$_2$O), which generates no ^{1}H signals. A wide variety of deuterated solvents are available, and Table 3-1 lists some representative ones. Such solvents are useful in another context, also. Many modern spectrometers use a deuterium lock system (Section 3-1A), and these solvents can also serve as the internal lock substance.

In the case of most other nuclei (e.g., ^{19}F, ^{31}P, etc.) it is quite easy to find solvents lacking the target nucleus. For ^{13}C, however, it is difficult. Certainly most organic solvents must, by their very nature, contain carbon. If solubility considerations require a carbon-containing solvent for a given ^{13}C experiment, it is best to pick a substance that gives as few ^{13}C signals as possible. Several common candidates are also listed in Table 3-1.

TABLE 3-1 Useful NMR Solvents

Solvent	Nucleus to be examined		
	^{1}H	^{13}C	Other
CCl$_4$	X	X	X
CH$_3$OH			X
CD$_3$OD	X	X	
CHCl$_3$			X
CDCl$_3$	X	X	X
H$_2$O		X	X
D$_2$O	X	X	X
CS$_2$	X	X	X
CF$_3$OD	X		
acetone-D$_6$	X		X
SO$_2$	X	X	X

Before we can go any further, we must be able to look at a molecular structure and decide how many signals of each type we can expect. To do this, we'll need a little training in symmetry, and that is our next topic.

SUMMARY

1. A typical nmr spectrometer consists of the following components:

   (a) A magnet capable of sustaining a strong, stable, homogeneous magnetic field. Magnets are classified as either permanent magnets, electromagnets, or superconducting magnets. The field strength of an electromagnet is controlled by electronic locking.
   (b) A probe within the magnetic field made up of a sample cavity, transmitter coil, and receiver coil.
   (c) Appropriate electronic circuitry, a computer, and peripheral devices to detect, amplify, and display the nmr signals.

2. Continuous wave (cw) nmr spectrometers are designed to operate in one of two ways:

   (a) Frequency sweep, where the magnetic field strength is fixed and the irradiation frequency is varied (swept).
   (b) Field sweep, where the irradiation frequency is fixed and the field strength is varied by means of sweep coils.

3. The signal-to-noise ratio of an nmr signal improves with $\sqrt{n}$, where n is the number of times the signal is measured and averaged.

4. The pulsed Fourier transform (PFT) technique for acquiring an nmr spectrum involves the following steps, once the sample is immersed in the magnetic field:

   (a) Irradiation of the sample with a short but powerful burst of rf radiation. The duration of the pulse (t_p) determines the range of frequencies covered.
   (b) Digital acquisition (monitoring) of the free induction decay (FID) signal induced in the receiver

coil as a function of time. The length of time between acquisition of each data point is called dwell time [t_d, Eq. (3-6)]. The total length of time that the FID signal is monitored is called the acquisition time (t_{acq}).

(c) If signal-to-noise considerations require FID signal averaging, multiple pulse sequences are carried out. There is a delay (t_w) within each pulse sequence to allow the nuclei to relax back to spin equilibrium (at a rate governed by T_1).

(d) Fourier transformation of the FID (time-domain) signal into a frequency-domain spectrum [Eq. (3-7)].

5. The line width (measured as $v_{1/2}$) of an nmr signal is a function of T_2^*, the effective spin–spin relaxation time, which in turn is mainly dependent on inhomogeneities in the magnetic field.

6. A given nmr spectrometer can only be configured to examine one isotope at a time, resulting in a homonuclear spectrum.

7. When preparing solution samples for nmr, care should be taken in the choice of a solvent. It is best if the solvent lacks any nuclei of the target isotope.

4 A LITTLE BIT OF SYMMETRY

THIS CHAPTER IS ABOUT

☑ **Symmetry Operations and Distinguishability**
☑ **Conformations and Their Symmetry**
☑ **Homotopic, Enantiotopic, and Diastereotopic Nuclei**
☑ **Accidental Equivalence**

4-1. Symmetry Operations and Distinguishability

Before we can begin to understand and predict the appearance of homonuclear nmr spectra, we must be able to recognize when the nuclei (and atoms) in a given structure will be distinguishable and when they will not. The test of distinguishability is based on symmetry relations among the nuclei, and it is these we will now explore.

A **symmetry operation** is defined as some actual or imagined manipulation of an object that leaves it indistinguishable from the original object. Doing *nothing* to an object leaves it indistinguishable, but we needn't consider this trivial **identity operation** further. Symmetrical objects are said to possess **symmetry elements**, which are points, axes, or planes that help describe symmetry operations.

Imagine a perfect glass sphere. Assuming that we leave its center of mass undisplaced, how might we manipulate it and still leave it indistinguishable? For one thing, we could rotate it around any imagined axis passing through its center, and it would appear unchanged. Not only that, but we could also rotate it any number of degrees around such an axis, and it would still be unchanged. Such a symmetry element is called a C_∞ **rotational axis** for reasons that will become clear shortly. Moreover, there is an infinite number of such axes passing through the center of the sphere.

Here is another manipulation we could perform. Suppose we were to pass an imaginary mirror directly through the center of the sphere. Certainly the mirror image of the left half would be indistinguishable from the actual right half, and vice versa. Thus, a sphere is said to have a **plane of symmetry** (symbol σ), because one half of the object is the mirror image of the other half. In fact, the sphere has an infinite number of such mirror planes.

Clearly, most objects are not as symmetrical as a sphere, which has an infinite number of axes and planes of symmetry. Still, many objects, and most molecules, have at least some symmetry. Let's discuss a few examples.

Chloromethane, structure **4-1**, has a nearly tetrahedral structure, where the chlorine and hydrogen atoms form the corners of a pyramid (tetrahedron) with carbon at its center.

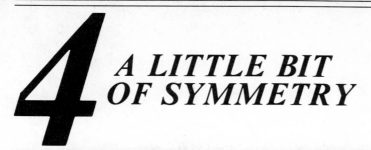

4-1

This molecule possesses several types of symmetry. Consider Figure 4-1. If you rotate the molecule around an imaginary axis passing through the carbon and chlorine atoms (colinear with the C—Cl bond), you will notice that every 120° rotation brings each hydrogen to a position previously occupied by

Side View　　　　　　　　　Top View

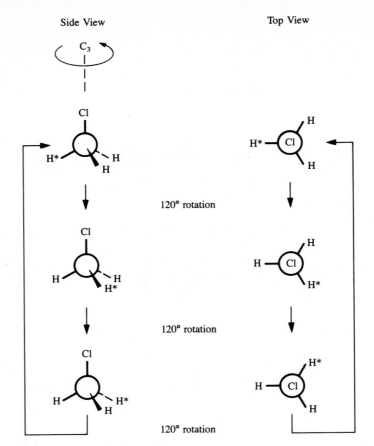

FIGURE 4-1 The C_3 axis of chloromethane. The circle represents the central carbon atom. One hydrogen is labeled (*) to show its position during rotation. It is actually indistinguishable from any other hydrogen.

another hydrogen. These hydrogens are said to be *related* by this operation and are, therefore, *indistinguishable* from one another. Notice that each 120° rotation provides an object indistinguishable from the original. Of course, after the third 120° rotation, we would return to where we started. And that is why this is called a C_3 **axis**: The object becomes indistinguishable from the original three times during one complete revolution (360°).

- An object with a C_n **axis** becomes indistinguishable n times during one 360° rotation (once every 360°/n degrees).

Notice also that if we were to rotate by any angle other than a multiple of 120°, the molecule would be *distinguishable* from the original. For example, rotation of structure **4-1** around its C_3 axis by 60° or 180° gives structures **4-1′** and **4-1″**, respectively. Do you see how they differ from structure **4-1**?

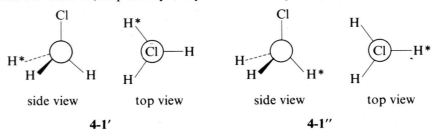

side view　　　　top view　　　　　　side view　　　　top view

4-1′　　　　　　　　　　　　　　**4-1″**

EXAMPLE 4-1 A certain object has a C_6 axis. (**a**) How many times does it become indistinguishable from the original during one full rotation? (**b**) How many degrees of rotation bring it to an indistinguishable position?

Solution:

(a) From the definition of a C_n axis, this object becomes indistinguishable 6 times during one 360° rotation around the axis.

(b) $360°/6 = 60°$. (Now you see why the axes of a sphere are C_∞. The sphere becomes indistinguishable an infinite number of times during one complete rotation.)

Reflection is also a symmetry operation. You may have realized that chloromethane also has three *planes* of symmetry, one for each C—H bond (and each containing the chlorine atom), as shown in Figure 4-2. Reflection in any of these σ planes relates one hydrogen to another, so mirror symmetry also tells us that the hydrogens are indistinguishable.

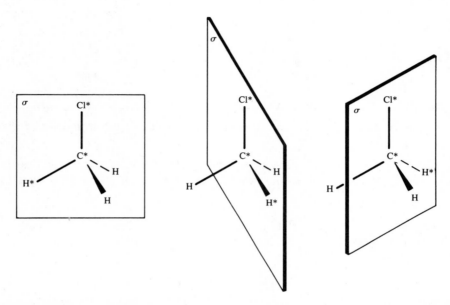

FIGURE 4-2 The mirror planes of chloromethane. The (*) indicates which atoms lie *in* the plane; the remaining hydrogens are related *by* the plane.

Here is the most important thing to remember:

- Atoms (i.e., nuclei) that are related by virtue of a symmetry operation are said to be *symmetry-* (or *chemically*) *equivalent*, and they are *isochronous*; that is, they *precess at exactly the same frequency*. Thus, symmetry-equivalent nuclei *cannot* be distinguished by nmr.

Therefore, the three hydrogens in chloromethane are symmetry-equivalent by virtue of both the C_3 axis and the symmetry planes and give only one signal in an ${}^1H$ nmr spectrum.

Next, consider the three isomers of dichloroethylene, structures **4-2**, **4-3**, and **4-4**.

$$\underset{\textbf{4-2}}{\overset{Cl}{\underset{H}{\diagdown}} C=C \overset{Cl}{\underset{H}{\diagup}}} \qquad \underset{\textbf{4-3}}{\overset{Cl}{\underset{H}{\diagdown}} C=C \overset{H}{\underset{Cl}{\diagup}}} \qquad \underset{\textbf{4-4}}{\overset{Cl}{\underset{Cl}{\diagdown}} C=C \overset{H}{\underset{H}{\diagup}}}$$

Each of these structures is *planar*; that is, all six atoms constituting each molecule lie in the same plane, called the molecular plane. What symmetry relationships can you find in these structures? Figure 4-3 shows that, in all three cases, the pairs of like atoms (H, C, and Cl) are symmetry-equivalent by virtue of C_2 axes and/or σ planes. Note also that for each of these three planar structures the molecule plane constitutes a plane of symmetry. But this symmetry element doesn't relate any of the atoms to another.

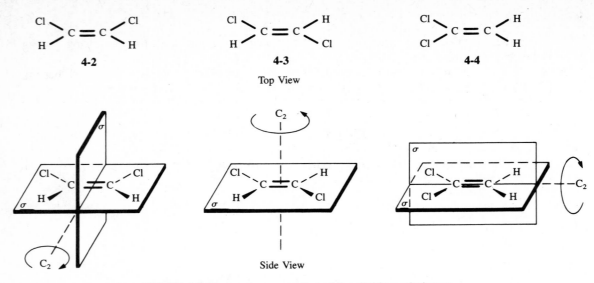

Top View

Side View

FIGURE 4-3 Symmetry properties of the dichloroethylenes.

EXAMPLE 4-2 How many ^1H, ^{13}C, and ^{35}Cl signals would you expect from each of the isomers in Figure 4-3?

Solution: In each case the two hydrogens are symmetry-equivalent, as are the two carbons and the two chlorines. Therefore, each isomer would exhibit a single ^1H signal, a single ^{13}C signal, and a single ^{35}Cl signal.

EXAMPLE 4-3 (a) What symmetry relationships can you discover for the *planar* molecule benzene, structure **4-5**? (b) How many ^1H and ^{13}C signals would you predict for benzene?

4-5

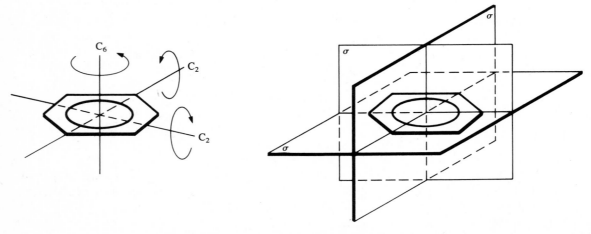

FIGURE 4-4 Some symmetry elements of benzene.

Solution:

    (a) All six hydrogens, as well as all six carbons, are symmetry-equivalent by virtue of a C_6 axis perpendicular to the ring and passing through its center. Additionally, the molecule has 6 C_2 axes in the plane of the ring, 6 σ planes perpendicular to the ring, and one more σ *in* the plane of the ring. *Some* of these are shown in Figure 4-4. Can you find the rest?

    (b) Because of the equivalences, benzene exhibits one ^1H signal and one ^{13}C signal.

4-2. Conformations and Their Symmetry

In Chapter 5 we'll look at the ^1H and ^{13}C nmr spectra of toluene (**3-1**, Section 3-3). Now, examine the two structures of toluene shown in Figure 4-5. These are referred to as **conformational structures** (or **conformations**, for short), because they are related by simple rotations of the ring-to-CH_3 bond, a process that is very facile. (There is actually an infinite number of such conformations, differing only in the angle of rotation around this bond.) The difference between these two is that in structure **B** one of the methyl (CH_3) hydrogens is *in* the plane containing the benzene (or phenyl) ring, while in structure **A**, one of the methyl hydrogens is in a plane *exactly perpendicular* to the ring. Inspection will reveal that structure **A** has a σ plane making two (of the three) methyl hydrogens equivalent to each other, the two ortho hydrogens (and carbons) equivalent to each other, and the two meta hydrogens (and carbons) equivalent to each other. On this basis, we would expect structure **A** to exhibit *five* carbon signals (the methyl carbon and the four sets of phenyl ring carbons) and *five* hydrogen signals (two types of methyl hydrogens and three sets of ring hydrogens). While our guess for the carbons and hydrogens on the ring turns out to be correct, we find that all three methyl hydrogens prove to be equivalent. Why is this so? The answer lies in the rate at which the ring–methyl bond rotates. Because this process is extremely rapid on the nmr time scale (see Section 1-4), the methyl group behaves as if it had a "local" C_3 axis, rendering all three hydrogens symmetry-equivalent on the time average. But remember: this is true only because of the high rate of rotation. If this rate were slow on the nmr time scale, we might expect to see multiple ^1H signals from the methyl group hydrogens.

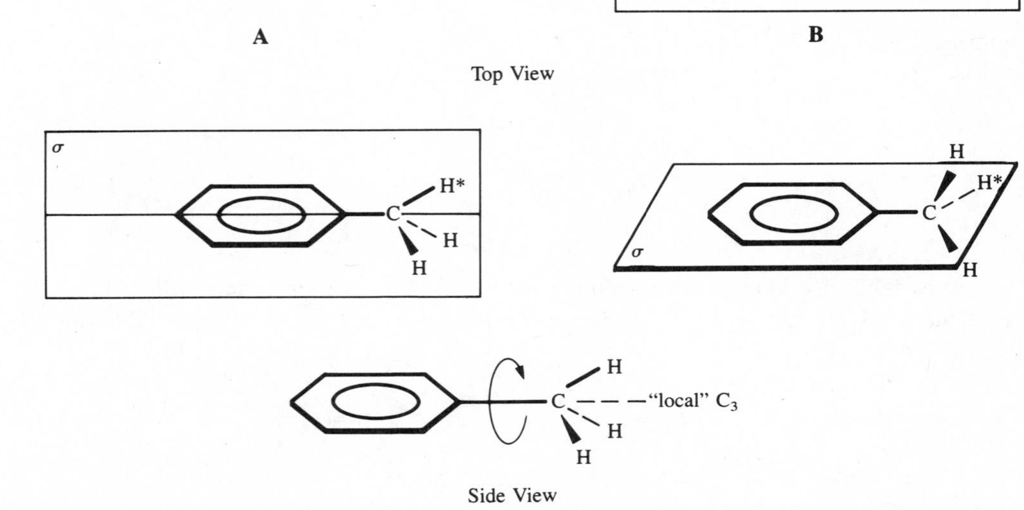

FIGURE 4-5 The symmetry properties of two conformations of toluene. H* indicates a hydrogen *in* the mirror plane; o, m, and p indicate the ortho, meta, and para positions.

EXAMPLE 4-4 What symmetry element(s) can you find in structure **B**, Figure 4-5?

Solution: There is a σ plane containing all six ring carbons, the five ring hydrogens, the methyl carbon, and *one* methyl hydrogen. The two remaining methyl hydrogens are related by this plane and are thus equivalent. Note that toluene *lacks* the C_6 axis present in benzene (Example 4-3).

- In principle, different conformations of a molecule exhibit different nmr spectral properties. However, if the conformations interconvert rapidly enough, they generate only an "averaged" spectrum.

In order to observe the separate spectra it is usually necessary to slow down the rotation of bonds in the molecule, and this is done by cooling the sample to a very low temperature. There will be more about these dynamic processes in Section 12-2.

4-3. Homotopic, Enantiotopic, and Diastereotopic Nuclei

Nuclei that are equivalent by virtue of a rotational axis are said to be *homotopic*. All of the equivalent nuclei discussed so far fall into this category. A set of homotopic nuclei gives only one nmr signal.

    Consider now the structure of bromochloromethane, structure **4-6**. Careful inspection will reveal the presence of a σ plane containing the Br, C, and Cl atoms and bisecting the H—C—H angle.

4-6

Therefore, these two hydrogens are symmetry-equivalent since they are related by reflection in the mirror plane. But because they are *not* related by any *axis* of symmetry, they are *not* homotopic. They are instead referred to as *enantiotopic* hydrogens. If you have difficulty deciding whether two nuclei are related by a mirror plane, another test for whether or not two atoms are enantiotopically related is called the **isotope substitution test**. To make the test, mentally substitute first one, then the other, suspected atom with a different isotope, and compare the two resulting structures. If the two structures are **enantiomers** (i.e., nonsuperimposable mirror images), the two suspected nuclei are said to be enantiotopically related. In the case of structure **4-6** the two substituted structures would be **4-6A** and **4-6B**. These two structures are, in fact, nonsuperimposable (as are your left and right hands), though you may wish to verify this by constructing molecular models of each one. We can summarize by saying that all enantiotopic nuclei are symmetry-equivalent (by virtue of a σ plane), but not all symmetry-equivalent nuclei are enantiotopic. To prove this, carry out the isotope substitution test on the equivalent nuclei in structures **3-1** and **4-1** through **4-5** and confirm that no two are enantiotopic.

    4-6A   **4-6B**

Structures **4-7** and **4-8** are more challenging examples. Are the two circled hydrogens equivalent or not? For structure **4-7** (in the conformation shown) there is a σ plane (can you find it?) that relates the two

    4-7   **4-8**

circled hydrogens. Application of the isotope substitution rule should also convince you that they are enantiotopic. But that same σ plane is *absent* in structure **4-8** because the carbon bearing the asterisk has four *different* substituents attached to it. In structure **4-7** the analogous carbon has only *three* different groups (the two methyls, like the two circled hydrogens, are enantiotopic and therefore equivalent). A tetrahedral atom with four different attached groups is referred to as *chiral* (or *asymmetric*). Applying the isotope substitution test to structure **4-8**, we generate structures **4-8A** and **4-8B**, which are *not* enantiomers, but rather are **diastereomers** (stereoisomers that are *not* enantiomers). In this case, the two circled hydrogens in **4-8** are said to be *diastereotopically related*, or simply *diastereotopic*.

4-8A   **4-8B**

And here is the all-important bottom line:

- Nmr normally cannot distinguish between enantiotopic nuclei, but it can (at least in principle) distinguish diastereotopic nuclei.

EXAMPLE 4-5 **(a)** Indicate all homotopic nuclei, enantiotopic nuclei, and diastereotopic nuclei in structure **4-9**. **(b)** How many carbon and hydrogen nmr signals would you predict for the compound? You may assume rapid rotation of all bonds.

4-9

Solution:

(a) First, notice that the carbon marked with an asterisk is chiral. The three hydrogens within each of the three methyl groups are equivalent by virtue of the rapid rotation of the C—C bonds (local C_3 axes). But the two methyls attached to the same carbon are *diastereotopic* (and therefore distinguishable), and each is also distinguishable from the third methyl. The remaining nuclei are all unique and therefore distinguishable.

(b) Each of the three methyls will exhibit its own carbon signal and hydrogen signal. In addition, the three remaining carbons and two remaining hydrogens will each give a separate signal, for a total of six carbon and five hydrogen signals.

4-4. Accidental Equivalence

At the end of Chapter 3 we intimated that the true value of nmr comes from its ability to distinguish *identical* nuclei (e.g., ^1H) that have *nonidentical* molecular environments. And, as stated in the previous section, we can in many cases even distinguish two hydrogens attached to the same carbon if they are diastereotopic. However, it occasionally happens that two nuclei that are *not* symmetry-equivalent in any way *accidentally* precess at exactly the same frequency and, hence, give rise to a single nmr signal. The two classic examples of this are propyne (structure **4-10**) and cyclopentanone (structure **4-11**). Both compounds would be expected to exhibit two hydrogen signals, yet both (at 60 MHz) display only one!

$$H_3C—C\equiv C—H$$

4-10

4-11

EXAMPLE 4-6 (a) Verify that structure **4-11** would be expected to show only two hydrogen signals. (b) How many carbon signals are expected?

Solution:

(a) The four hydrogens closer to the C=O form a symmetry-equivalent set by virtue of two σ planes and a C_2 axis (can you find these symmetry elements?). The same is true of the two carbons to which these hydrogens are attached. The four remaining hydrogens form a second set of equivalent hydrogens, and the carbons to which they're attached form a second set of equivalent carbons. Thus, there are only *two* hydrogen signals expected.

(b) The two sets of carbons, plus the C=O carbon, give rise to *three* carbon signals.

It turns out that accidental equivalence is most common in ^{1}H nmr, and much less prevalent in the case of ^{13}C.

Now that we have an idea of what to expect, let's proceed to the next chapter and examine some actual nmr spectra.

SUMMARY

1. A symmetry operation is the manipulation of an object in such a way that it is indistinguishable from the original object. There are two principal types of symmetry elements, axes and planes.

2. An *n*-fold axis of symmetry (designated C_n) is an imaginary axis passing through the center of the object. Rotation around this axis by $360°/n$ leaves the object indistinguishable from the original.

3. A plane of symmetry (designated σ) is an imaginary (mirror) plane through the center of an object. The half of the object on one side of the plane is the mirror image of the half on the other side.

4. Conformations of a molecule are the possible arrangements of the atoms in the molecule that preserve the molecule's chemical bonds. One conformation of a molecule differs from any other by the degree of rotation around its bonds.

5. Nuclei that are related by one or more symmetry elements are indistinguishable by nmr methods. They are said to be symmetry- or chemically equivalent.

6. Nuclei that are related by an axis of symmetry are called homotopic.

7. Nuclei that are related only by a plane of symmetry are called enantiotopic.

8. Nuclei that are *not* symmetry-equivalent but nonetheless precess at the same frequency are called accidentally equivalent.

5 SOME TYPICAL NMR SPECTRA

THIS CHAPTER IS ABOUT

☑ **The ¹H NMR Spectra of Toluene at 60 and 80 MHz**
☑ **The Chemical Shift Scale**
☑ **The 300-MHz ¹H NMR Spectrum of Toluene**
☑ **The ¹³C NMR Spectra of Toluene at 20 and 75 MHz**
☑ **Data Acquisition Parameters**

5-1. The ¹H NMR Spectra of Toluene at 60 and 80 MHz

The ultimate value of any type of spectroscopy depends on our ability to interpret accurately the spectral data we acquire. These data are usually plotted as a spectrum, which is nothing more than a graph of signal intensity versus frequency (also commonly called *position*) for each signal. In this chapter we will examine and interpret the actual ¹H and ¹³C spectra of toluene.

A. The 60-MHz ¹H spectrum

From our discussion (Chapter 4) of the symmetry properties of toluene (structure **5-1**), we decided that toluene should exhibit four ¹H signals: one for the three equivalent methyl hydrogens, one for the two equivalent ortho ring hydrogens (H_o), one for the two equivalent meta ring hydrogens (H_m), and one for the para ring hydrogen (H_p).

5-1

Figure 5-1 shows as actual ¹H nmr spectrum of toluene, recorded at an operating frequency of 60.0 MHz and a field strength of 1.41 T. Not only does it show three signals instead of four, but the signal at "zero" (marked with an asterisk) is actually from another source to be discussed later. Thus, toluene itself exhibits only *two* ¹H signals at 60 MHz, one at 140 Hz downfield (to the left) of "zero," and a somewhat more intense one at 430 Hz downfield of "zero." The Hz scale is shown above the abscissa. There are several other observations we can make. From the slight "ringing" pattern (Section 3-4A) on the signal at "zero" we can deduce that the spectrum was obtained under continuous-wave conditions, by scanning from left to right. The S/N ratio (Section 3-2C) is > 36, and the spinning side bands (Section 3-2C) are too small to be observed.

Recall from Section 3-1A that even minuscule changes in magnetic field strength cause observable changes in precession frequency. It is, therefore, virtually impossible to measure *absolute* frequencies with the high degree of precision and reproducibility necessary in nmr. However, it *is* possible to measure frequency *differences* very precisely. For example, suppose we *were* able to measure frequencies to a precision level of ± 1 Hz in 60 MHz (1 part in 60 million). Our first nmr measurement for toluene might show the two signals at 60,000,140 and 60,000,430 Hz. But our next trial might produce values of 59,999,527 and 59,999,817 Hz, respectively, because of a small drift in

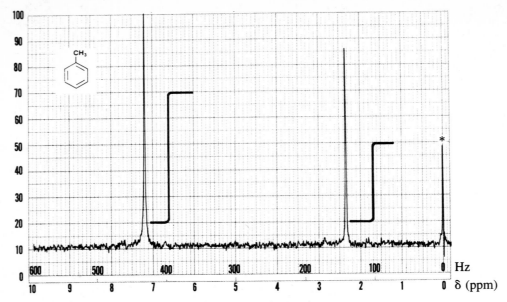

FIGURE 5-1 The 60-MHz ^{1}H nmr spectrum of toluene (from *The Aldrich Library of NMR Spectra*).

field strength. Notice, however, that the *difference* in position of the two peaks is *exactly* 290 Hz in both experiments.

Recognizing that there is this slight unavoidable drift in absolute frequencies but a constancy of frequency *differences*, how might we devise a reproducible scale for the abscissa of an nmr spectrum? A little thought should suggest the answer. We include in each sample a small amount (~ 1–5% by volume) of a **reference compound** that gives rise to a sharp signal somewhat apart from the other signals of interest. Although the reference signal will drift along with the signals from the sample, the *difference* in frequency between the reference signal and any other signal is always constant. We will *arbitrarily* assign the reference signal the value zero and measure frequency *differences* in Hz downfield or upfield (to the right) of the reference signal.

Perhaps this discussion of reference compounds may remind you of the lock substances mentioned in Section 3-1A. Indeed, in some older continuous-wave spectrometers, the reference signal *was* used as the lock signal. There were two major problems with this, however. First, it required large amounts of the reference compound to be dissolved in the sample, and second, it was very difficult to observe signals located near the lock signal because of beat interference patterns. In modern spectrometers this problem is usually avoided through the use of a **heteronuclear lock**, that is, a lock substance that contains only nuclei different from the nuclei that will generate the signals of interest. Thus, a deuterium (^{2}H) lock is generally used in the case of ^{1}H or ^{13}C nmr.

Because we can only examine one type of nucleus at a time (Section 3-4B), the *reference* signal must come from the *same* type of nucleus as the sample, e.g., an ^{1}H reference signal for ^{1}H nmr, a ^{13}C reference signal for ^{13}C nmr, etc. The compound should be chemically inert, soluble in a variety of solvents, and readily removable if the sample is to be reclaimed. A perfect candidate in the case of ^{1}H and ^{13}C spectroscopy is **tetramethylsilane** (TMS, structure **5-2**) in which all twelve hydrogens are symmetry-equivalent (why?), as are the four carbons. The use of TMS as an internal reference is so widespread that you may assume its signal defines the zero point in all ^{1}H and ^{13}C spectra. Reference compounds for other nuclei are more varied, and some of these are listed in Table 5-1. It is always a good practice to make certain what the reference compound is before attempting to interpret an nmr spectrum.

$$CH_3$$
$$H_3C—Si—CH_3$$
$$CH_3$$

5-2

TABLE 5-1
NMR Reference Compounds

Nucleus	Reference compound
^{1}H	$(C\underline{H}_3)_4Si$ (TMS)
^{2}H (D)	$(CH_3)_3(C\underline{D}H_2)Si^a$
^{13}C	$(\underline{C}H_3)_4Si$ (TMS)
^{15}N	anh. $\underline{N}H_3$ (external)
^{19}F	$CCl_3\underline{F}$
^{31}P	85% $H_3\underline{P}O_4$ (external)

^{a} Natural abundance 0.2% in normal TMS.

Now that we've agreed on an abscissa scale (Hz up- or downfield from TMS, in the case of ^{1}H and ^{13}C), let's consider the scale for intensity. It is fairly obvious that the signal at 430 Hz in Figure 5-1 is taller than the one at 140 Hz. But is signal height alone the best measure of intensity? Look at the three signals in Figure 5-2. Although each one has a different height, the actual *intensities* of the signals are equal. This is because signal intensity is measured not by signal height alone, but by *area* under the curve, which is roughly equal to the peak height times half-width (Section 3-4A).

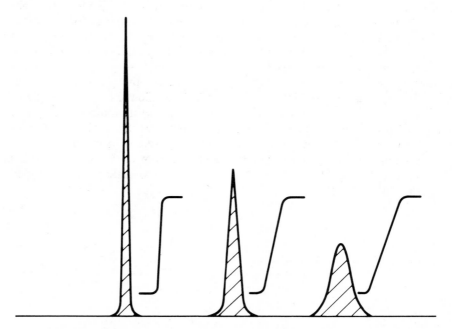

FIGURE 5-2 Three peaks of different height and width but equal area.

EXAMPLE 5-1 Using a ruler, measure the peak height and half-width of each signal in Figure 5-2 and confirm that all three are equally intense.

Solution: Measure the half-width (Section 3-4A) of each peak and multiply this by the height of the peak. The exact numbers you calculate will depend on the scale of your ruler, but the three products should be essentially identical.

Look again at the 60-MHz spectrum of toluene. Can you measure half-widths accurately enough to decide if the downfield signal is *really* more intense? In general, this is difficult to do, especially for sharp signals. But don't despair! All nmr spectrometers can electronically *integrate* (i.e., find the relative area under) any peak in the spectrum. This integral is shown by the S-shaped line beside each signal in Figures 5-1 and 5-2. The *vertical* displacement of the "integral" line is proportional to signal area and, hence, intensity.

The integrals in Figure 5-1 tell us that the ratio of intensity of the downfield signal to that of the upfield signal is 5:3. This is very significant, because (all other things being equal)

- ^1H nmr signal intensity is directly proportional to the number of ^1H nuclei giving rise to that signal.

Since the integral ratio for toluene is 5:3 and there are eight hydrogens in the molecule, we can infer that the downfield signal represents five hydrogens, and the upfield signal three. But which hydrogens go with which signals?

Because toluene exhibits fewer ^1H signals than we expected on the basis of symmetry considerations (Chapter 4), there must be accidental equivalence (Section 4-4) of some of the nuclei. Which ones would you expect to be the most similar? Judging from the structure of toluene and the integral ratios, it would be a reasonable guess (and is indeed correct!) to assign the downfield signal to the five accidentally equivalent hydrogens attached to the phenyl ring, and the upfield signal to the three equivalent methyl hydrogens.

B. The 80-MHz ^1H spectrum

Figure 5-3 shows the ^1H nmr spectrum of toluene, obtained this time on an instrument operating at 80.0 MHz and at a field strength of 1.88 T. This spectrum has somewhat sharper spectral signals and no ringing, evidence that it was obtained by the PFT technique (Section 3-3). If you look closely, you'll notice several other differences. First, the downfield signal is no longer obviously more intense than the upfield one, although the integral still indicates a 5:3 ratio. Because of the way the frequency domain spectrum is reconstructed after the Fourier transformation (Section 3-3), the relative intensities of signals are often not readily apparent by visual inspection. It is always best to depend on the electronic integration for exact relative intensities.

But there's another much more significant difference: the position of the peaks. Notice that the smaller signal is now 187 Hz downfield of TMS while the larger one is 573 Hz downfield of TMS. A moment's reflection (and a review of Section 2-2A) should convince you that this is exactly in accord with our nmr theory. Equation (2-4) indicates that *differences* in precessional frequencies (as well as the frequencies themselves) are directly proportional to field strength. Since the spectrometer's operating frequency increased from 60 MHz to 80 MHz (and its field strength from 1.41 T to 1.88 T), the separation between any two peaks is also expanded by one third $(140 \rightarrow 187$ Hz; $430 \rightarrow 573$ Hz).

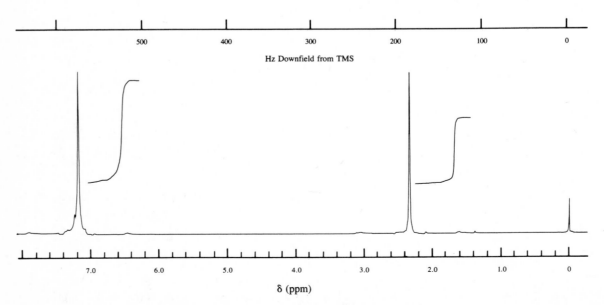

Hz Downfield from TMS

δ (ppm)

FIGURE 5-3 The 80-MHz ^1H nmr spectrum of toluene.

5-2. The Chemical Shift Scale

A problem is now apparent. Even though we all agree to use TMS as our reference signal, the spectral data acquired from your 60-MHz instrument are different from my 80-MHz spectrum of the same compound. Isn't there a system we could both adopt that would give the same abscissa scale for both spectra? Indeed, there is! Since the frequency *difference* ($\Delta v_i = v_{signal} - v_{reference}$) between any signal and the reference signal is directly proportional to operating frequency v_o [and field strength—don't forget Eq. (3-1)] we will define a new quantity called **chemical shift** (δ) by the equation

DELTA CHEMICAL SHIFT SCALE
$$\delta_i = \frac{\Delta v_i}{v_o} \times 10^6 = \frac{\Delta v_i}{v_o'} \tag{5-1}$$

where v_o carries units of Hz, and v_o' is in MHz. The δ scale is actually dimensionless, but its "units" are usually expressed as ppm (parts per million) as a consequence of the factor of 10^6 following the dimensionless ratio in Eq. (5-1).

EXAMPLE 5-2 Calculate the chemical shifts of the two toluene signals, using both the 60-MHz and 80-MHz data, and confirm that both sets of data yield identical values of chemical shift for each signal.

Solution: For the *upfield* signal at 60 MHz

$$\delta = \frac{\Delta v}{v_o'} = \frac{140 \text{ Hz}}{60 \text{ MHz}} = 2.33 \text{ ppm}$$

For the *downfield* signal at 60 MHz

$$\delta = \frac{\Delta v}{v_o'} = \frac{430 \text{ Hz}}{60 \text{ MHz}} = 7.17 \text{ ppm}$$

At 80 MHz the two signals occur at 187 Hz and 573 Hz, respectively. Thus,

$$\delta_{upfield} = \frac{187 \text{ Hz}}{80 \text{ MHz}} = 2.33 \text{ ppm}$$

$$\delta_{downfield} = \frac{573 \text{ Hz}}{80 \text{ MHz}} = 7.17 \text{ ppm}$$

Voilà!

EXAMPLE 5-3 The 60-MHz ^1H nmr spectrum of methyllithium (CH_3Li) shows a signal at 126 Hz *upfield* of TMS. What is its chemical shift?

Solution: Recall that $\Delta v_i = v_{signal} - v_{reference}$. A signal *upfield* of TMS is defined as having a *negative* value of v_{signal}. Thus,

$$\delta = \frac{\Delta v}{v_o'} = \frac{-126 \text{ Hz}}{60 \text{ MHz}} = -2.10 \text{ ppm}$$

Therefore Eq. (5-1) requires that signals *downfield* from the reference signal have *positive* chemical shifts, and that those *upfield* of the reference signal have *negative* chemical shifts. In the older literature there was some confusion about whether the upfield or downfield signals should carry positive chemical shifts, but the above convention has since been universally adopted.

Now look back at Figures 5-1 and 5-3. In addition to the Hz scale, the δ scale is shown along the bottom of each abscissa.

note: Equation (5-1) also requires that 1 δ unit equals 60 Hz at 60 MHz, 80 Hz at 80 MHz, etc.

At one time an alternate ^1H chemical shift scale called the **tau** (τ) **scale** was used, defined by the relationship

TAU CHEMICAL SHIFT SCALE
$$\tau = 10 - \delta = 10 - \frac{\Delta v_i}{v_o'} \tag{5-2}$$

EXAMPLE 5-4 Calculate the chemical shifts in τ units for the signals in Examples 5-2 and 5-3.

Solution: Substitute the values of δ from Examples 5-2 and 5-3 into Eq. (5-2). For toluene

$$\tau_{\text{upfield}} = 10 - \delta_{\text{upfield}} = 10 - 2.33 = 7.67 \text{ ppm}$$

$$\tau_{\text{downfield}} = 10 - \delta_{\text{downfield}} = 10 - 7.17 = 2.83 \text{ ppm}$$

For methyllithium

$$\tau = 10 - \delta = 10 - (-2.10) = 12.10 \text{ ppm}$$

note: Although the τ scale is not used much any more, you may find it sometimes in older literature. So beware!

5-3. The 300-MHz ^{1}H NMR Spectrum of Toluene

By using the high field of a superconducting magnet, as well as the PFT technique, we can generate the 300-MHz ^{1}H spectrum of toluene, as shown in Figure 5-4. In this figure, only the δ scale is shown.

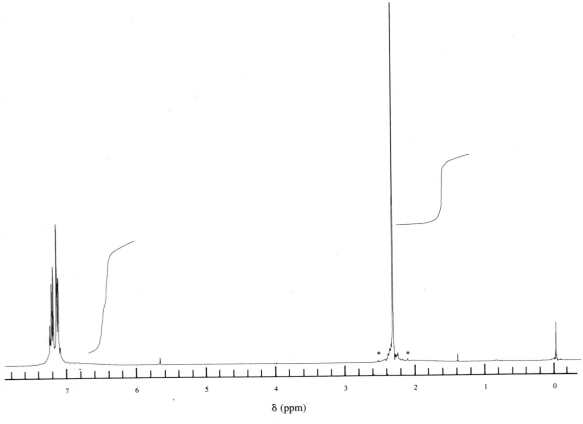

δ (ppm)

FIGURE 5-4 The 300-MHz ^{1}H nmr spectrum of toluene.

EXAMPLE 5-5 (a) What is the field strength needed to generate a ^{1}H frequency of 300 MHz? **(b)** What is the position of the methyl signal (Figure 5-4) in Hz downfield of TMS? **(c)** The spinning side bands for the methyl signal are marked with asterisks. Estimate the spinning rate.

Solution:

(a) As in Example 2-7, solve Eq. (2-6) for B_0.

$$B_0 = \frac{2\pi v}{\gamma} = \frac{(2\pi \text{ rad})(300 \times 10^6 \text{ s}^{-1})}{267.5 \times 10^6 \text{ rad T}^{-1}\text{s}^{-1}} = 7.05 \text{ T}$$

(b) Even before we look at Figure 5-4 we know the methyl signal has a chemical shift of 2.33 ppm (Example 5-2). Rearranging Eq. (5-1) gives us

$$\Delta v = \delta(v_o') = (2.33 \text{ ppm})(300 \text{ MHz}) = 700 \text{ Hz}$$

(c) At 300 MHz, each δ unit equals 300 Hz, and each tick on the abscissa denotes 60 Hz. The spacing between the sideband and main peak is therefore approximately 60 Hz, and this equals the spin rate (see Example 3-5).

The integration again indicates a 5:3 ratio of phenyl ring hydrogens to methyl hydrogens, but do you notice anything different? Rather than being a single peak as in Figures 5-1 and 5-3, the ring hydrogen signal is now split into at least *six* signals. This is because we are operating at such a high field strength that signals that were previously accidentally equivalent are now capable of being resolved (separated). You may, in fact, have already noticed that the downfield signal in Figures 5-1 and 5-3 was broader, especially at its base, than the methyl signal. The lesson to be gained from this section is that

- The higher the operating field (and frequency) of an instrument, the better resolution it is able to provide.

By the way, if you're wondering why there are at least six ring hydrogen peaks, rather than just the three we expected on the basis of symmetry, you'll find out in Chapter 8!

5-4. The ^{13}C NMR Spectrum of Toluene at 20 and 75 MHz

A. The 20-MHz ^{13}C spectrum

Figure 5-5 is a ^{13}C nmr spectrum of toluene, obtained by the PFT technique on the same instrument with the same magnetic field strength that provided the 80-MHz ^{1}H spectrum in Figure 5-3.

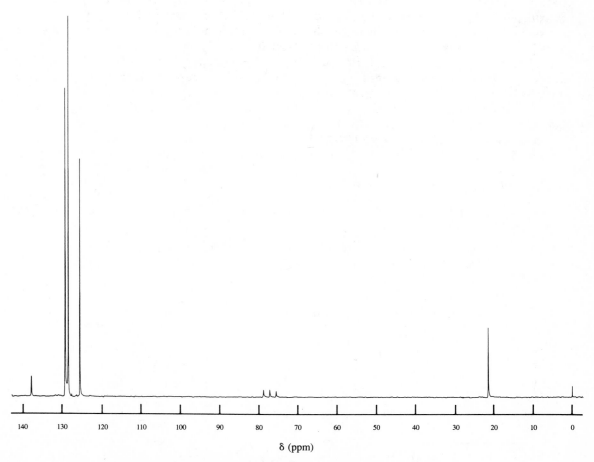

δ (ppm)

FIGURE 5-5 The 20-MHz ^{13}C nmr spectrum of toluene.

EXAMPLE 5-6 **(a)** What operating frequency was used to obtain the ^{13}C spectrum? **(b)** How many Hz correspond to each δ unit?

Solution:

(a) The field strength that generates a ^{1}H frequency of 80 MHz is 1.88 T [from Section 5-1B or by solving Eq. (2-6) for B_0]. To find the frequency for the ^{13}C nuclei, solve Eq. (2-6) for v, and substitute for γ the value for ^{13}C from Table 2-1.

$$v = \frac{\gamma B_0}{2\pi} = \frac{(67.264 \times 10^6 \text{ rad T}^{-1}\text{s}^{-1})(1.88 \text{ T})}{2\pi \text{ rad}} = 20.1 \text{ MHz}$$

(b) From Eq. (5-1), if $v_o = 20.1$ MHz, each δ unit equals 20.1 Hz.

Once again TMS (its *carbon* signal) is used to define "zero" on the chemical shift scale. Furthermore, the three signals from δ 75 to 80 ppm arise from the solvent (CDCl$_3$), which also provides a deuterium lock signal. The reason CDCl$_3$ exhibits three carbon signals, rather than just one, will be explained in Section 8-6B. Besides those peaks, there are five others at δ 21.4, 125.6, 128.5, 129.2, and 137.6 ppm, exactly in accord with our expectations (Section 4-1). Because of the similarity of this spectrum to the previously discussed ^{1}H spectra, it seems reasonable (and is indeed correct!) to assign the upfield signal to the methyl *carbon*, and the four relatively close signals to the four nonequivalent phenyl ring *carbons* (Section 4-1).

But wait! If the above assignments are correct, shouldn't the intensity ratios of these five signals be (from right to left) 1:1:2:2:1? The answer is yes, and no. It turns out there are several factors that control the *relative* intensity of nmr signals generated by the PFT technique. The most important of these are relative relaxation times and the *nuclear Overhauser effect*. The dependence on relaxation time of relative signal intensity is far more significant with the PFT method than with the continuous-wave method. This is because the PFT technique interprets the signal from a rapidly relaxing nucleus as more intense than the signal from a slowly relaxing nucleus, since the rapidly relaxing nuclei return more quickly to a Boltzmann distribution (Section 2-3B). In the case of ^{1}H nuclei, the span of relaxation times is fairly small, and therefore this factor is roughly the same for all ^{1}H nuclei in a molecule. With ^{13}C nuclei, however, there can be large differences in relaxation times and thus a significant span of intensities for equal numbers of nonequivalent carbons. In general, the more hydrogens attached to a carbon, the faster the carbon can relax, and the more intense is its PFT signal. One way to circumvent this problem is to provide long delay times (Section 3-3) between pulses, but because of time constraints this is not always practical.

The **nuclear Overhauser effect (NOE)** is the enhancement of intensity of an nmr signal generated by one nucleus when it is near another nonequivalent nucleus simultaneously being irradiated. We will discuss this effect in more detail later (Chapter 11), but for now, here is its most significant result:

- The more hydrogens attached to a given carbon, the more intense is its signal (all other things being equal).

This fact helps us to assign the phenyl ring carbon signals to the various positions in toluene. The weakest signal (at δ 137.6 ppm) must correspond to the carbon lacking hydrogens, C$_1$ in struc-

TABLE 5.2 The Data Acquisition Parameters for the Toluene PFT Spectra[a]

	^{1}H (80 MHz)	^{1}H (300 MHz)	^{13}C (20 MHz)	^{13}C (76 MHz)
n (number of pulses)	8	8	2000	620
t_p (pulse duration), μs	5	12	6	16
t_{acq} (acquisition time), s	3	1.35	0.77	0.204
t_d (dwell time), μs	367	166	94	25
t_w (delay time), s	3	3	15	15

[a] Refer to Figure 3-13.

ture **5-3**. Each of the remaining ring carbons has a single hydrogen attached, but there are two equivalent C_2's, two equivalent C_3's, and only one C_4. The weaker signal at 125.6 ppm can therefore be assigned to C_4. The remaining two signals are harder to assign, especially since their relative intensity depends on the exact parameters of the pulse sequence (see Table 5-2). We will learn much more in Chapter 6 about assigning such signals, but it turns out that the δ 128.5 ppm signal results from C_3, while the δ 129.2 ppm signal is from C_2. Empirically, we notice that the closer a ring carbon is to the methyl group, the farther downfield its signal occurs.

5-3

EXAMPLE 5-7 (a) How many signals do you expect to see in the ^{13}C nmr spectrum of cyclohexanone, structure **5-4**? (b) Which signal should be the *least* intense?

5-4

Solution:

(a) Symmetry considerations (Chapter 4) lead us to conclude that there are four sets of carbons:

(b) Of the four signals, the carbonyl (C=O) signal will be weakest, for the carbonyl carbon lacks any hydrogens attached directly to it (see Problem 6 in Self-Test I).

There is another very significant difference between the ^1H spectra we saw previously and this ^{13}C spectrum. Did you notice it? The ^1H signals of toluene differ in chemical shift by only about 5 ppm, which would correspond to 100 Hz at 20 MHz. The ^{13}C signals, on the other hand, occupy a span of nearly 120 ppm (2400 Hz)! In fact, the ^1H signals of most known compounds show up in a fairly narrow range of chemical shift, about δ 0 to 10 ppm (1000 Hz at 100 MHz), while ^{13}C signals span a range of $\sim$250 ppm (5000 Hz at 20 MHz). These spans, in fact, determine the spectral width discussed in Section 3-3. For this reason, the chance of accidental equivalence is far *smaller* in the case of ^{13}C spectra than with ^1H spectra.

B. The 75-MHz ^{13}C spectrum

Before looking at Figure 5-6, try to guess the effect on the ^{13}C spectrum of toluene of increasing the field strength by a factor of 3.75. Although the signal positions will spread out (when measured in Hz), the chemical shifts (δ) should remain unaffected. Relative intensities might vary a little because of different pulse parameters. But, by and large, the spectrum should not change significantly, since there are no accidental equivalences to resolve. Now look at Figure 5-6. You're right—no significant change! The relative intensities of the ring carbon signals have changed a little, but we would have no difficulty in recognizing this to be a ^{13}C spectrum of toluene.

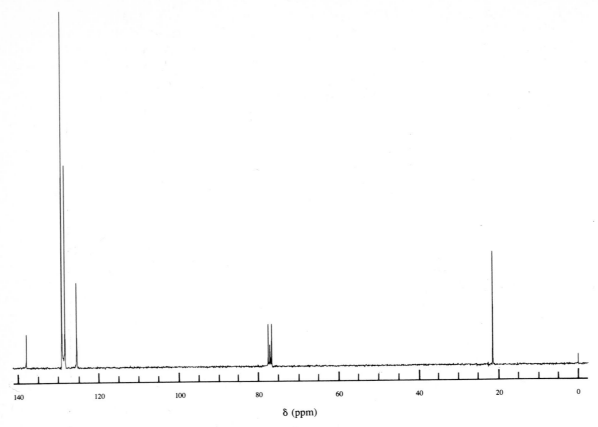

FIGURE 5-6 The 75-MHz ^{13}C nmr spectrum of toluene.

5-5. Data Acquisition Parameters

As mentioned in Section 3-3, the success of the PFT technique depends on the careful selection of data acquisition parameters. These data for the spectra shown in Figures 5-3–5-6 appear in Table 5-2. The same sample was used in all four cases. Notice the following trends. First, as expected, it takes many more pulses to generate an adequate ^{13}C spectrum than a ^{1}H spectrum because of the lower sensitivity of the carbon nuclei (Section 2-2A). The pulse duration increases with increasing field strength. Because the carbon spectral width is substantially larger (see Sections 3-3 and 5-4A) than that of hydrogen, lower resolution can be tolerated in ^{13}C spectra than in ^{1}H spectra. Therefore, the dwell time is generally *shorter* for carbon, and since this fills the microprocessor's memory faster, acquisition times are shorter for carbon than for hydrogen. Finally, longer delay times between pulses are required for carbon because of the larger span of relaxation times. Lest you worry that you're going to have to provide values for all these parameters to generate PFT nmr spectra, relax! Most modern spectrometers have menu programs that automatically select the appropriate values in most cases. Moreover, these parameters are of only very limited importance for *interpreting* the resulting spectra.

Before we leave the discussion of these ^{13}C spectra, there is one more qualifier to add. Both of these ^{13}C spectra involved the use of a technique known as ^{1}H-decoupling. This technique will be described more fully in Chapter 11, but for now suffice it to say that if we had *not* made use of ^{1}H-decoupling, these spectra would have looked *far* more complicated.

By now, we have seen and interpreted several actual ^{1}H and ^{13}C spectra over a large span of frequency and field strength. We have made some empirical observations about factors that influence the position and intensity of the various signals. For example, there appears to be a rough correlation between the *relative* chemical shifts of ^{1}H nuclei and the relative chemical shifts of the carbons to which they're attached. In the next two chapters, our goal is to systematize a very large body of nmr data in order to be able to assign nmr signals to specific nuclei within any molecule and to predict where such signals might occur even before seeing the spectrum!

SUMMARY

1. Positions of nmr signals are measured in hertz (Hz) from some standard reference signal. The reference signal comes from a compound, usually added to the sample, called the reference compound. For ^1H and ^{13}C nmr, the reference compound is tetramethylsilane (TMS).
2. The intensity of an nmr signal is directly proportional to the area (rather than peak height) of the signal. This area is measured by electronically integrating the signal.
3. So that the position of nmr signals can be reported independently from the spectrometer's operating frequency (and field strength), positions are usually expressed in terms of chemical shift (δ), defined by

$$\delta_i = \frac{\Delta v_i}{v_o'} \tag{5-1}$$

where v_o' is the operating frequency of the instrument in MHz.

4. The intensity of ^1H nmr signals is directly proportional to the number of hydrogens giving rise to that signal. The intensity of ^{13}C signals is normally *not* directly proportional to the number of carbons giving rise to the signal. This is because different carbons usually have a greater range of relaxation times and because of NOE enhancements.
5. The higher the operating frequency (and field) of an nmr spectrometer, the better the resolution it provides in nmr spectra.

6 CORRELATING PROTON CHEMICAL SHIFTS WITH MOLECULAR STRUCTURE

THIS CHAPTER IS ABOUT

☑ **Shielding and Deshielding**
☑ **Methyl Hydrogen Chemical Shift Correlations**
☑ **Methylene Hydrogen Chemical Shift Correlations**
☑ **Vinyl Hydrogen Chemical Shift Correlations**
☑ **Magnetic Anisotropy**
☑ **Aromatic Hydrogen Chemical Shift Correlations**
☑ **Hydrogens Attached to Elements Other than Carbon**

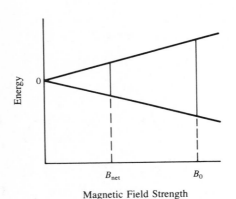

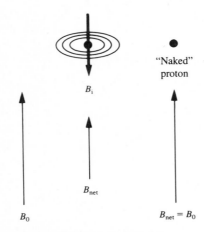

FIGURE 6-1 Effect of diamagnetic shielding.

6-1. Shielding and Deshielding

From our discussion in Chapter 5, we can begin to appreciate the ability of nmr spectroscopy to act as a sensitive probe of molecular structure. We saw how the ^{1}H nmr spectra of toluene readily distinguish the methyl hydrogens from the ring hydrogens (and even the individual ring hydrogens from one another at very high field) and how the ^{13}C spectra distinguish the carbons from one another. In this chapter we shall explore semi-quantitative relationships between the chemical shift of a given ^{1}H nucleus and its location in a molecular structure.

Let's return for a moment to the 60-MHz ^{1}H nmr spectrum of toluene, Figure 5-1. In it we saw three signals: TMS (reference) hydrogens at δ 0.00 ppm, methyl hydrogens at 2.33 ppm, and ring hydrogens at 7.17 ppm. Suppose you were asked to indicate which of these hydrogens precesses with the highest frequency at a given field strength. If you recall from Section 3-2 that frequency increases *from right to left* (or that field strength increases from left to right) in a typical spectrum, you can see that the TMS hydrogens precess the slowest, while the ring hydrogens precess the fastest *at a given field strength*. Conversely, we could say that *at constant frequency*, the TMS hydrogen signals occur at the highest field (i.e., the TMS hydrogens require the highest field to enter resonance) while the ring hydrogen signals occur at the lowest field. It turns out that if we could obtain an nmr signal for a "naked" ^{1}H nucleus (a proton free of all electrons and other molecular entanglements), its signal would appear far downfield of even the toluene ring hydrogens. Why does a "naked" proton enter resonance at such a low field while ^{1}H nuclei within molecules (and surrounded by electrons) require a much higher field?

The answer lies in a very simple fact: The electron cloud surrounding each nucleus in a molecule serves to *shield* that nucleus from the applied magnetic field. Let's see how this happens. Figure 6-1 depicts a comparison between a "naked" proton and one shielded by an electron cloud. Just as the oscillating nuclear magnetic field passing through the receiver coil generates an electric current (Section 2-3), a current moving through a wire generates a magnetic field (as in electromagnets, Section 3-1). In an exactly analogous way, the applied magnetic field (B_0 in Figure 6-1) causes the electrons surrounding the nucleus to circulate through their orbitals in such a way as to generate an induced magnetic field (B_i) *in opposition to the applied field*. As a result, while the "naked" proton experiences the full magnitude of the applied field, the shielded nucleus experiences a *net* (or effective) field (B_{net}) which is equal to the applied field minus the induced field.

NET MAGNETIC FIELD
$$B_{net} = B_0 - B_i \qquad \textbf{(6-1)}$$

Because the strength of the induced field is directly proportional to that of the applied field, we can define a **shielding constant** σ (which is a function of the molecular environment of the nucleus) by the equation

INDUCED MAGNETIC FIELD
$$B_i = \sigma B_0 \qquad \textbf{(6-2)}$$

Next, Eq. (6-1) can be recast in the form

NET MAGNETIC FIELD
$$B_{net} = B_0(1 - \sigma) \qquad \textbf{(6-3)}$$

Substituting this result into Eq. (2-6) gives

PRECESSION FREQUENCY
$$\nu_{precession} = \left(\frac{\gamma B_0}{2\pi}\right)(1 - \sigma) \qquad \textbf{(6-4)}$$

Thus, the greater the shielding of the nucleus, the lower its resonance frequency and the farther to the right it will appear in an nmr spectrum. Conversely, nuclei from which electron density has been withdrawn are said to be *deshielded* and appear toward the left of the spectrum (higher frequency).

The type of shielding and deshielding we've been discussing so far is a consequence of the spherically symmetric (Chapter 4) circulation of electrons and is referred to as **diamagnetic shielding**. The diamagnetic shielding constant is proportional to electron density around the nucleus. However, there may also be *asymmetric* circulation of the electrons because of the structure of the molecule (e.g., if it has unpaired electrons), and this results in a *paramagnetic* component of the shielding. **Paramagnetic shielding**, it turns out, is negative in the sense that the field induced by paramagnetic shielding is aligned *with* (rather than opposed to) the applied field.

EXAMPLE 6-1 Does paramagnetic shielding cause a nucleus to enter resonance at higher or lower field?

Solution: Referring to Figure 6-1, we note that if B_i is aligned *with* B_0, then B_{net} will equal $B_0 + B_i$. This *larger* net field will cause the nucleus to precess at a higher frequency (or at a lower applied field).

In the case of hydrogen (which has its electrons in spherically symmetric orbitals), diamagnetic shielding is the predominant term. But for other nuclei (e.g., carbon, fluorine, phosphorus, etc.) paramagnetic shielding (which has the opposite effect) becomes increasingly important. It is possible, by means of sophisticated molecular orbital calculations, to estimate the electron distribution within molecules and thereby to make quantitative guesses about the amount of shielding at each nucleus. However, we will adopt a semi-empirical approach, trying to discern from large amounts of spectral data the likely position of signals for a variety of nuclei.

6-2. Methyl Hydrogen Chemical Shift Correlations

So far we have discussed the ^1H chemical shifts of the methyl (CH_3) groups in toluene (δ 2.33 ppm), TMS (0.00 ppm), and CH_3Li (-2.10 ppm, Example 5-3), as well as of the ring hydrogens of toluene (7.17 ppm). In Figure 6-2, the relative positions of these signals are depicted graphically. Of these, the CH_3Li has the hydrogens most shielded, owing to the electron-donating effect of the lithium atom. The methyl

FIGURE 6-2 Partial ^{1}H chemical shift range.

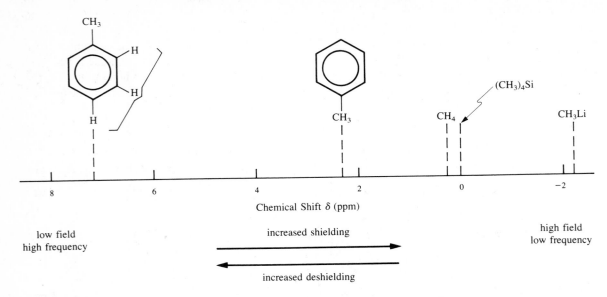

hydrogens in TMS are the next most shielded, and this shielding can be attributed to the electron-rich silicon atom nearby. The methyl group in toluene is somewhat less shielded (or more deshielded), and the ring hydrogens are the most deshielded of all, for reasons we will explain in Section 6-6. One way to begin to organize such data, at least for the methyl groups, is to regard each of these compounds as a derivative of methane, CH_4, whose ^{1}H signal occurs at δ 0.23 ppm. Next, we can define a **substituent shielding parameter** ($\Delta\delta_X$) for substituent X by the equation

SHIELDING PARAMETER $$\Delta\delta_X = \delta_{CH_3X} - \delta_{CH_4} = \delta_{CH_3X} - 0.23 \qquad \textbf{(6-5)}$$

EXAMPLE 6-2 Calculate $\Delta\delta_X$ for the following X groups: Li, Si(CH$_3$)$_3$, H, and phenyl (C$_6$H$_5$).

Solution: Using Eq. (6-5),

$$\Delta\delta_{Li} = -2.10 - 0.23 = -2.33$$

In like manner, the $\Delta\delta$ values for Si(CH$_3$)$_3$, H, and phenyl are -0.23, 0.00, and 2.10, respectively.

Notice from Example 6-2 that a large *positive* value of $\Delta\delta$ corresponds to a *downfield* shift (resulting from *deshielding* of the nucleus) and that a *negative* value corresponds to an *upfield* shift (from *shielding* of the nucleus), relative to the effect of X = H. For this reason, it might seem more logical to call $\Delta\delta_X$ a *deshielding* parameter, but we will follow the convention and refer to it as the shielding parameter of group X.

EXAMPLE 6-3 Derive the relationship between the shielding constant σ (Eqs. 6-2 to 6-4) and the substituent shielding parameter $\Delta\delta$. [*Hint:* you'll need to remember Eq. (5-1).]

Solution: From Eq. (6-5)

$$\Delta\delta_X = \delta_{CH_3X} - \delta_{CH_4}$$

Using Eq. (5-1), we find

$$\Delta\delta_X = \frac{\nu_{CH_3X} - \nu_{TMS}}{\nu'_o} - \frac{\nu_{CH_4} - \nu_{TMS}}{\nu'_o} = \frac{\nu_{CH_3X} - \nu_{CH_4}}{\nu'_o}$$

Applying Eq. (6-4) gives us

$$\Delta\delta_X = \frac{\dfrac{\gamma B_0}{2\pi}[(1-\sigma_X)-(1-\sigma_H)]}{v_0'} = \frac{\gamma B_0}{2\pi v_0'}(\sigma_H - \sigma_X)$$

Thus, a substituent that is more strongly *shielding* than H ($\sigma_X > \sigma_H$) has a *negative* value of $\Delta\delta$ (corresponding to an upfield shift), while a less shielding substituent ($\sigma_X < \sigma_H$) has a *positive* $\Delta\delta$ (corresponding to a downfield shift).

When ^{1}H nmr spectroscopy was in its infancy, J. W. Shoolery began analyzing the spectra of literally hundreds of compounds in an attempt to extract average values of $\Delta\delta_X$ for a wide variety of common substituents. These values have been refined by him and others, and the concept has been extended to ^{13}C chemical shifts as well as to those of some other common nuclei. Table 6-1 lists some of the most commonly encountered substituent groups and their average shielding parameters. Notice that the *deshielding* ability of a substituent *increases* as you go *down* the table. Note also that the tabulated value for X = phenyl (1.85) is slightly different from the value we calculated in Example 6-2 (2.10). This is because the latter value comes from a specific compound (toluene), while the former value represents an average value from many phenyl-containing molecules.

EXAMPLE 6-4 Using the data in Table 6-1, together with Eq. (6.5), estimate the chemical shift of the methyl hydrogens in the following compounds:

$$\text{(a) } CH_3I \qquad \text{(b) } CH_3-\overset{\overset{\textstyle O}{\|}}{C}-CH_3 \qquad \text{(c) } CH_3-O-\overset{\overset{\textstyle O}{\|}}{C}-CH_3 .$$

Solution:

(a) Solve Eq. (6-5) for δ_{CH_3X} and substitute the value from Table 6-1, $\Delta\delta_I = 1.82$ ppm, for $\Delta\delta_X$.

$$\delta_{CH_3I} = \Delta\delta_I + 0.23 = 1.82 + 0.23 = 2.05 \text{ ppm}$$

The actual value from the spectrum of CH_3I is 2.15 ppm.[1]

(b) By symmetry (Chapter 4), both methyl groups in this molecule are equivalent. When you use the $\Delta\delta$ value for $-C\overset{\textstyle O}{\underset{\textstyle R}{\diagdown}}$, you find that both methyl groups should appear at $1.70 + 0.23 = $ 1.93 ppm. The observed value is 2.17 ppm.[1]

(c) The two methyl groups in this structure are *not* equivalent; one is attached to an oxygen atom, the other to a carbon atom. The chemical shift of the former is calculated from the $\Delta\delta$ value for the $-O-C\overset{\textstyle O}{\underset{\textstyle R}{\diagdown}}$ group (3.13), while the latter requires the value for the $-C\overset{\textstyle O}{\underset{\textstyle OR}{\diagdown}}$ group (1.55). Thus, the calculated chemical shifts are δ 3.36 and 1.78 ppm, respectively.

$$CH_3-O-\overset{\overset{\textstyle O}{\|}}{C}-CH_3$$

	3.36	1.78
calculated	3.36	1.78
observed[2]	3.71	2.09

EXAMPLE 6-5 A certain molecule has molecular formula C_2H_3N but unknown structure. It exhibits a single peak in its ^{1}H nmr spectrum at δ 2.05 ppm.[2] Propose a structure for the molecule.

Solution: Because there is only one peak, all three of the hydrogens are equivalent, probably occurring as a methyl group. This would leave one carbon and one nitrogen as "X" to account for the observed chemical shift. Notice in Table 6-1 that a $-C\equiv N$ group has a $\Delta\delta$ value of 1.70, so a methyl attached to a $-CN$ group should occur near $1.70 + 0.23 = 1.93$ ppm. Indeed, CH_3CN *is* the suspect molecule.[1]

TABLE 6-1 Shielding Parameters for Common Substituent Groups[a,b]

X	$\Delta\delta_X$ (ppm)	X	$\Delta\delta_X$ (ppm)	X	$\Delta\delta_X$ (ppm)
—CH_3 (or R)	0.47	—C(=O)R	1.70	—O—R	2.36
—C=C<	1.32	—C≡N	1.70	—Cl	2.53
—C≡C—	1.44	—I	1.82	—O—H	2.56
—C(=O)OR	1.55	—C(=O)Ph	1.84	—NR_3^+	2.70
—NR_2	1.57	—Ph	1.85	—O—C(=O)R	3.13
—C(=O)NR_2	1.59	—NH—C(=O)R	2.27	—O—SO_2R	3.13
—S—R	1.64	—Br	2.33	—O—Ph	3.23
				—NO_2	3.80
				—F	4.00

[a] These data are from Dailey, B. P., and Shoolery, J. W., *J. Am. Chem. Soc.* **77**: 3977, 1955, as quoted in reference 3 at the end of this chapter.
[b] R represents any alkyl group, consisting solely of carbons and hydrogens singly bonded to one another; Ph represents phenyl (C_6H_5).

In most cases the use of these shielding parameters gives a reasonable estimate (± 0.2 ppm) of the expected chemical shift of a given hydrogen. Once you've tentatively identified the compound from its spectrum, you can look up the actual spectrum of the assigned structure in references (such as those listed at the end of this chapter) to confirm the identification.

6-3. Methylene Hydrogen Chemical Shift Correlations

A. Noncyclic molecules

A CH_2 group within a molecule is referred to as a **methylene group**. Suppose such a group were connected to two of the substituent groups from Table 6-1. Could you predict the chemical shift of the methylene hydrogens? If the two substituents (X and Y) exert their shielding effects independently, then perhaps the chemical shift of the methylene group could be calculated by simply adding the shielding parameters of both substituents to the chemical shift of methane.

METHYLENE CHEMICAL SHIFT $\delta_{X—CH_2—Y} = 0.23 + \Delta\delta_X + \Delta\delta_Y$ **(6-6)**

Let's see how well this **additivity principle** works.

EXAMPLE 6-6 Calculate the chemical shift of the methylene hydrogens in

(a) Cl—CH_2—Cl   (b) Br—CH_2—C(=O)—Ph   (c) Ph—CH_2—C(=O)—CH_3

Solution:

(a) By substituting the shielding parameter for Cl (2.53) into Eq. (6-6) for *both* X and Y, we calculate

$$\delta_{Cl—CH_2—Cl} = 0.23 + \Delta\delta_{Cl} + \Delta\delta_{Cl} = 0.23 + 2.53 + 2.53 = 5.29 \text{ ppm}$$

The observed chemical shift of the hydrogens in this compound is 5.30 ppm.[1]

(b) Similarly,

$$\delta_{Br—CH_2—C(=O)—Ph} = 0.23 + \Delta\delta_{Br} + \Delta\delta_{—C(=O)Ph} = 0.23 + 2.33 + 1.84 = 4.40 \text{ ppm}$$

The observed value is 4.44 ppm.[2]

(c) Remember, we're after the methylene chemical shift, so

$$\delta_{\mathrm{Ph-CH_2-\overset{O}{\overset{\|}{C}}-CH_3}} = 0.23 + \Delta\delta_{\mathrm{Ph}} + \Delta\delta_{-\mathrm{C}\overset{\nearrow O}{\underset{R}{\searrow}}} = 0.23 + 1.85 + 1.70 = 3.78 \text{ ppm}$$

The actual value is 3.67 ppm.[2]

We see from this example that the additivity principle works very well indeed. You might wonder if such an approach can be extended to **methine groups** (a carbon connected to *one* hydrogen and three other substituent groups). Unfortunately, in such cases the three substituent groups tend to interfere with each other so that their shielding effects are *not* exerted independently.

EXAMPLE 6-7 Predict the chemical shift of the *methine* hydrogen in

$$
\begin{array}{c}
\mathrm{O-CH_2-CH_3} \\
| \\
\mathrm{H-C-O-CH_2-CH_3} \\
| \\
\mathrm{O-CH_2-CH_3}
\end{array}
$$

Solution:

$$\delta_{\mathrm{C-H}} = 0.23 + 3(\Delta\delta_{\mathrm{OR}}) = 0.23 + 3(2.36) = 7.31 \text{ ppm}$$

The value actually observed is 5.16 ppm.[2], so the additivity principle failed us in this case.

B. Cyclic molecules

Many chemical compounds contain rings formed by the cyclic connection of three or more atoms. Toluene (Chapters 3, 4, and 5), for example, possesses a ring composed of six carbons. Let's see if you can use the logic developed so far to estimate the chemical shifts of some cyclic compounds.

EXAMPLE 6-8 Predict the chemical shift of the hydrogens in the following cyclic structures. For the purposes of this problem you may assume that the rings are planar, though actually they are not.

Solution:

(a) By symmetry (a C_5 axis), all five of the methylene groups are equivalent. We can estimate their chemical shift by treating them as $R-CH_2-R$ groups, where $\Delta\delta_R$ is 0.47 (Table 6-1). Thus

$$\delta_{\mathrm{CH_2}} = 0.23 + 0.47 + 0.47 = 1.17 \text{ ppm}$$

The observed value is 1.51 ppm.[2]

(b) Using the same reasoning, we would predict that all *six* equivalent methylenes in this structure should appear at the same chemical shift, about 1.17 ppm. In actuality, this compound's signal occurs at 1.43 ppm.[1]

Example 6-8 tempts us to predict that all simple cyclic structures composed of three or more methylenes should exhibit ^{1}H nmr signals at about the same position. Although this is true for rings with five or more methylenes, the smaller rings show significant variations (see Table 6-2). The reason for the anomalous behavior of the small rings ($n = 3$ and 4) will be described in Section 6-5. But more

TABLE 6-2 ^{1}H Chemical Shifts
of Simple Cycloalkanes^a

(CH$_2$)$_n$

n	δ (ppm)
3	0.22
4	1.96
5	1.51
6	1.43
7	1.53
8	1.57
10	1.51

^{a} Data from references 2 and 3 at
the end of the chapter.

important for the present purpose is that we can use the chemical shifts in Table 6-2, rather than the value 0.23 (methane), as base values when estimating the chemical shifts of hydrogens in cyclic compounds.

EXAMPLE 6-9 Estimate the chemical shift of the underlined hydrogen in

Solution: Using a base value of δ 1.51 for a CH$_2$ group in a five-membered ring (Table 6-2), and the $\Delta\delta_X$ value for Cl (Table 6-1), we predict

$$\delta_{C-H} = 1.51 + 2.53 = 4.04 \text{ ppm}$$

The observed value is 4.37 ppm.[2] Not bad!

You may wonder what effect, if any, the chlorine has on the chemical shifts of the *other* hydrogens in Example 6-9. Not unexpectedly,

• The deshielding effect of a substituent normally falls off dramatically as the number of intervening bonds increases.

The chlorine is separated from the underlined hydrogen by *two* bonds; this is called a *geminal* relationship, and here the effect is greatest ($\Delta\delta = 2.53$ ppm). The hydrogens (H*) *three* bonds away from the chlorine (called a *vicinal* relationship) are deshielded only slightly (about 0.5 ppm from their normal position), while the deshielding of those hydrogens (H') separated by four bonds is negligible.

Before we proceed any further, there's something else that might be troubling you. Why do we need all these correlations and tables if we can simply look up the actual spectra of so many compounds? There are two answers to this question. First, in trying to identify an unknown compound we need to start somewhere, and its nmr spectrum is the best place. With only that, we can usually (if it's not too complicated a molecule) make a guess about its structure, then go to the literature to see if we're correct. But what if the spectrum of your compound has never been reported before? In that case our main evidence for its structure may be how well its spectrum matches expectations based on these correlations.

EXAMPLE 6-10 Propose a structure for the molecule whose 60-MHz ^{1}H nmr spectrum is reproduced in Figure 6-3, given only that its molecular formula is $C_6H_{12}O$.

note: As in the rest of this book, when the integral line (Section 5-1) for the intensity of a signal is not shown in a spectrum, the numerical value of the intensity is given at the top of the signal.

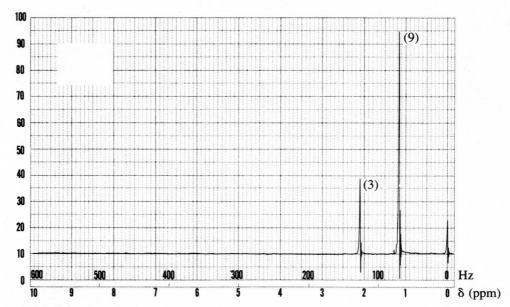

FIGURE 6-3 Unknown compound with molecular formula $C_6H_{12}O$ (from *The Aldrich Library of NMR Spectra*).

Solution: We note immediately that there are *nine* equivalent hydrogens at δ 1.16 ppm, and three other equivalent ones at δ 2.10 ppm. Whenever you encounter a nine-proton signal near δ 1.0 ppm, it is probably due to a tertiary butyl group [$(CH_3)_3C—$]. Note how all three of the methyls in such a group are equivalent, just as the three hydrogens in a methyl group are equivalent (Section 4-2). The δ 2.10-ppm signal fits very well for a methyl group next to a C=O group (see Table 6-1 and Example 6-4). The correct structure, therefore, is

6-4. Vinyl Hydrogen Chemical Shift Correlations

Hydrogens attached directly to carbons that are doubly bonded to other carbons are called **vinyl (or olefinic) hydrogens**.

EXAMPLE 6-11 Identify the vinyl hydrogens in

Solution:

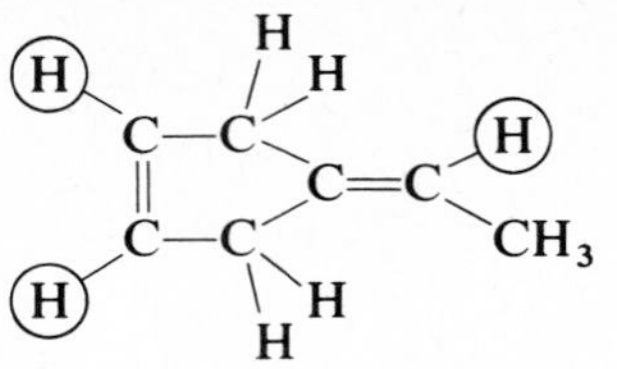

The orbitals with which a doubly bonded carbon forms bonds are more *electronegative* than the orbitals of singly bonded carbons. The more electronegative an orbital is, the more it tends to attract the bonding electrons toward itself. Vinyl hydrogens, therefore, have less electron density around them than, say, the hydrogens in methane. As a direct result (see Section 6-1), vinyl hydrogens are *deshielded*, and their signals appear downfield from those of hydrogens attached to singly bonded carbons. For example, ethylene (structure **6-1**) exhibits one ^{1}H signal (all four hydrogens are equivalent, right?) at δ 5.28 ppm. Vinyl hydrogens typically appear in the δ 4.5–7.0-ppm region of the ^{1}H spectrum.

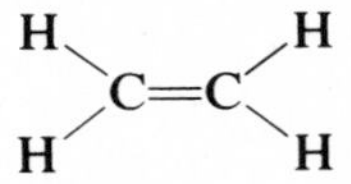

6-1

We can establish a list of shielding parameters for vinyl substituents (Table 6-3) similar to the one we had for methyl substituents (Table 6-1). Notice that the magnitude of a substituent's effect is strongly dependent on its location relative to the hydrogen. This time, we'll use δ 5.28 ppm (the chemical shift of ethylene) as our base value in the calculations.

TABLE 6-3 Vinyl Hydrogen Chemical Shift Correlations[a]

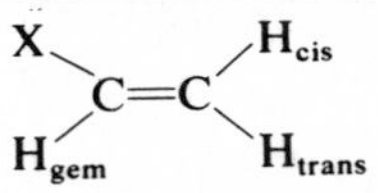

X	gem	cis	trans
—R	0.44	−0.26	−0.29
—CH$_2$—	0.67	−0.02	−0.07
—CH$_2$Cl	0.72	0.12	0.07
—C≡N	0.23	0.78	0.58
—C=C—	0.98	−0.04	−0.21
—C=O	1.10	1.13	0.81
—C(=O)OH	1.00	1.35	0.74
—C(=O)OR	0.84	1.15	0.56
—OR	1.18	−1.06	−1.28
—phenyl	1.35	0.37	−0.10
—Cl	1.00	0.19	0.03
—Br	1.04	0.40	0.55
—NR$_2$	0.69	−1.19	−1.31

[a] Data from Pascual, C., Meier, J., and Simon, W., *Helv. Chim. Acta* **49**: 164, 1966, as quoted in reference 3 at the end of this chapter. Recall that a *negative* value of $\Delta\delta$ corresponds to an *upfield* shift.

EXAMPLE 6-12 Predict the chemical shifts of all three vinyl hydrogens in styrene, structure **6-2**.

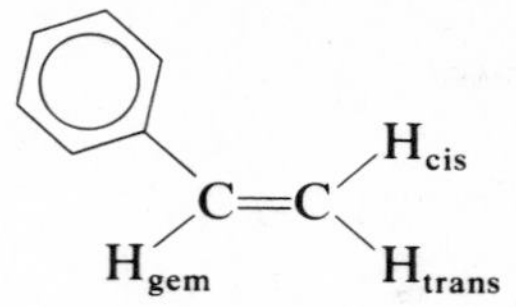

6-2

Solution: For each hydrogen, add the value of $\Delta\delta$ for a phenyl group to the base value of δ for a hydrogen in ethylene.

$$\delta_{gem} = 5.28 + 1.35 = 6.63 \text{ ppm} \quad (\text{found: } 6.66 \text{ ppm}^2)$$

$$\delta_{cis} = 5.28 + 0.37 = 5.65 \text{ ppm} \quad (\text{found: } 5.64 \text{ ppm}^2)$$

$$\delta_{trans} = 5.28 + (-0.10) = 5.18 \text{ ppm} \quad (\text{found: } 5.18 \text{ ppm}^2)$$

The close agreement of these numbers is no accident. Styrene was the model compound used to generate the $\Delta\delta$ values for a phenyl substituent! Now, let's try a much more challenging example.

EXAMPLE 6-13 Suggest a structure for the molecule whose molecular formula is $C_5H_8O_2$ and which exhibits ^1H nmr signals at δ 6.13 (1 H), 5.59 (1 H), 3.79 (3H), and 1.98 ppm (3H).

Solution: This problem requires us to use everything we've learned so far in this chapter. From the chemical shifts and integrals it looks as though this molecule possesses two nonequivalent vinyl hydrogens and two nonequivalent methyl groups. From Table 6-1, the methyl group at δ 3.79 is probably connected to a $-O-\overset{\displaystyle O}{\overset{\displaystyle \|}{C}}-R$ group $(3.79 - 0.23 = 3.56)$, suggesting the partial structure $H_3C-O-\overset{\displaystyle O}{\overset{\displaystyle \|}{C}}-R$, where R must consist of the remaining three carbons and five hydrogens. Here are several possible structures:

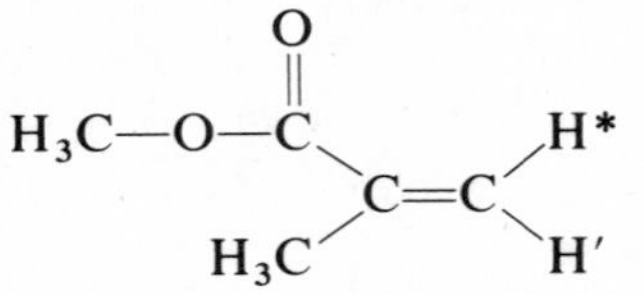

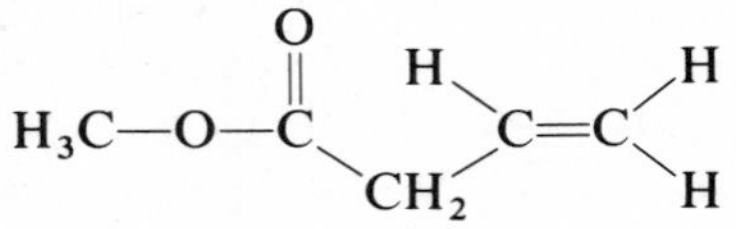

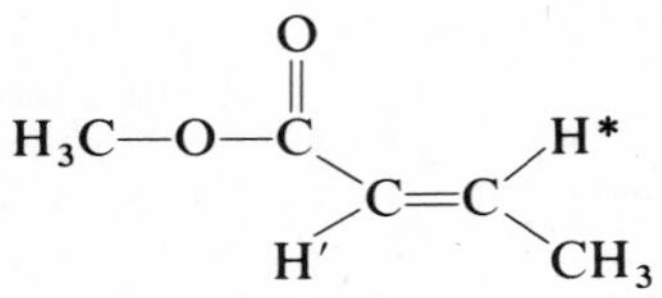

We can immediately rule out structure **B**, for it has three vinyl hydrogens and one methylene group and so would produce *four* additional signals. But how do we decide among the remaining three structures? Right! We use the data in Table 6-3 to calculate the expected chemical shifts of the two vinyl hydrogens in each structure. In **A**, H* is *cis* to the CO_2CH_3 group and *trans* to the CH_3 group, so if we apply the additivity principle

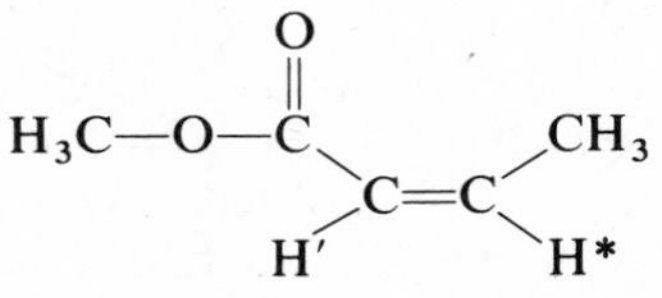

$$\delta_{H^*} = 5.28 + 1.15 + (-0.29) = 6.14 \text{ ppm}$$

Similarly,

$$\delta_{H'} = 5.28 + 0.56 + (-0.26) = 5.59 \text{ ppm}$$

Now carry out the same calculations for structures **C** and **D**. For **C**

$$\delta_{H*} = 5.28 + 1.15 + 0.44 = 6.87 \text{ ppm}$$

and

$$\delta_{H'} = 5.28 + 0.84 + (-0.26) = 5.86 \text{ ppm}$$

For **D**

$$\delta_{H*} = 5.28 + 0.44 + 0.56 = 6.28 \text{ ppm}$$

and

$$\delta_{H'} = 5.28 + 0.84 + (-0.29) = 5.83 \text{ ppm}$$

Which of the structures do you think fits best? The values for structure **A** obviously fit the best—and this result supports our extending the additivity principle to vinyl hydrogens.

There is another type of hydrogen that, at first glance, appears quite similar to a vinyl hydrogen. It is called an **aldehydic** (or formyl) **hydrogen** and is attached to a carbon that is doubly bonded to an oxygen.

$$\begin{array}{c} O \\ \parallel \\ C \\ H \diagup \quad \diagdown R \end{array}$$

The C=O group (called a **carbonyl group**) has a strong deshielding effect (see Tables 6-1 and 6-3). For this reason, aldehydic hydrogens are significantly more deshielded than even vinyl hydrogens. In fact, they are found in a remote region of the ^{1}H spectrum, around 9–10 ppm, where almost nothing else appears.

6-5. Magnetic Anisotropy

Knowing that a vinyl hydrogen is quite a bit more deshielded than typical methyl or methylene hydrogens, what would you predict about the properties of a hydrogen (called an **acetylenic hydrogen**) attached to a carbon triply bonded to another carbon (H—C≡C—)? Most people would predict such a hydrogen to be even more deshielded than a vinyl hydrogen. But they'd be wrong! Acetylenic hydrogens normally appear in the region around δ 2.0–3.0 ppm, significantly *upfield* of their vinyl cousins. Why is this so?

The currently accepted explanation of this anomaly involves the special way electrons in the triple bond behave when immersed in a magnetic field. Recall (Section 6-1) that when electrons are subjected to a strong applied magnetic field, they circulate in such a way as to induce a smaller magnetic field that opposes the applied field. An applied magnetic field causes the electrons in the triple bond to circulate around the internuclear (bond) axis, thereby inducing a field along the internuclear axis in opposition to the applied field. This is depicted in Figure 6-4, where the lines of magnetic flux are shown as dotted lines. Note that in the region near the internuclear axis, the induced field, being opposed to the applied field, will *shield* nuclei situated there. A hydrogen attached to a triply bonded carbon lies along this axis (see Table 9-1) and therefore enjoys the shielding effect of the induced field, causing a substantial *upfield* shift in the position of its nmr signal.

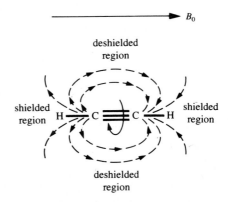

FIGURE 6-4 Anisotropic induced magnetic field (dotted lines) in the region around a triple bond.

Notice also that in the regions *perpendicular* to the triple bond axis, the induced field is aligned *with* the applied field. Because the direction and magnitude of the induced field vary as a function of position, we say the induced field is an **anisotropic field**.

EXAMPLE 6-14 Suppose a weird molecule could be created in which a hydrogen was forced to occupy a position midway between two triply bonded carbons but off the internuclear axis. What effect would its location have on the position of its nmr signal?

Solution: The induced field in the cylindrical region around the triple bond is aligned *with* the applied field (see Figure 6-4), so the *net* field (Section 6-1) experienced by a hydrogen in this region would be *greater* than the applied field alone. This would cause the signal to move downfield (or to higher frequency), the result of apparent deshielding. Believe it or not, such molecules (e.g., cyclic acetylenes) *have* been synthesized!

It is believed that similar anisotropic induced fields in the regions around small rings account for their anomalous chemical shifts (Section 6-3).

6-6. Aromatic Hydrogen Chemical Shift Correlations

Do you remember from our discussion in Chapter 5 that in toluene the five hydrogens attached to the ring appear at δ 7.17 ppm? These hydrogens actually belong to a special subclass of vinyl hydrogens, though this may not have been apparent from the structure shown in Section 3-3. A somewhat more informative structural drawing of toluene is

Molecules that possess a six-membered ring with alternating double and single bonds have very special chemical and physical properties. Compounds with such rings are referred to as **aromatic compounds**. This term has nothing to do with their fragrance. Instead, it refers to the unique characteristics of the cyclic network of alternating double and single bonds.

You may have noticed that the "aromatic" hydrogens (those attached to an aromatic ring) are more *deshielded* than typical vinyl hydrogens. This is a direct consequence of the magnetic anisotropy of the aromatic ring. As shown in Figure 6-5, the electrons of the alternating double bonds interact in such a way as to form a donut-shaped cloud of electrons above and below the plane of the ring. (This is why the structures of aromatic compounds are often shown with circles inside the ring; see Section 3-3 and Examples 4-3 and 4-4.) When immersed in an external magnetic field, these electrons begin to circulate just as the electrons in a triple bond do (Figure 6-4), generating an analogous induced field whose lines of magnetic flux are shown in Figure 6-5. Look carefully at the *direction* of the induced field. Directly above and below the center of the ring, the induced field is opposed to the applied field, giving rise to a *shielding* effect on nuclei located in that region. However, outside the periphery of the ring, the induced field is aligned *with* the applied field, causing *deshielding* of nuclei in that region. Now, where are the hydrogens attached to an aromatic ring located?

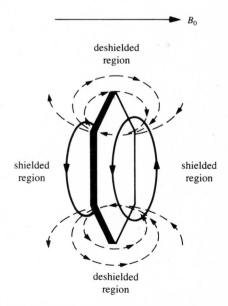

FIGURE 6-5 Anisotropic induced magnetic field (dotted lines) around an aromatic ring. Think of the ring as perpendicular to the page.

**TABLE 6-4 Aromatic Substituent
Shielding Parameters[a]**

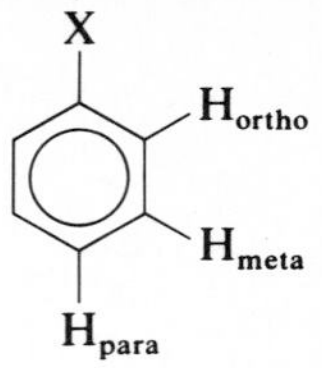

| | $\Delta\delta$ (ppm) | | |
X	ortho	meta	para
—NO₂	0.97	0.30	0.42
—C(=O)O—R	0.93	0.20	0.27
—C(=O)R	0.63	0.27	0.27
—Cl	0.0	0.0	0.0
—CH₃ (R)	−0.10	−0.10	−0.10
—OR	−0.23	−0.23	−0.23
—F	−0.32	0.00	−0.20
—OH	−0.37	−0.37	−0.37
—NH₂	−0.77	−0.13	−0.40
—NR₂	−0.50	−0.20	−0.50

[a] Data from Pasto, D. J. and Johnson, C. R., *Organic Structure Determination*, Prentice-Hall, Englewood Cliffs, New Jersey, 1969.

Around the periphery. As a result, they experience a *deshielding* effect from the circulation of the aromatic electrons.

Typical aromatic hydrogens usually occur in the δ 6.5–8.0-ppm chemical shift range. And just as for vinyl hydrogens, correlation tables of shielding parameters can be developed for aromatic substituents. Only in this case we use the chemical shift of benzene (δ 7.27 ppm) as our base value in the calculations. Table 6-4 lists several common aromatic substituent parameters.

EXAMPLE 6-15 Predict the chemical shift of the aromatic hydrogens in

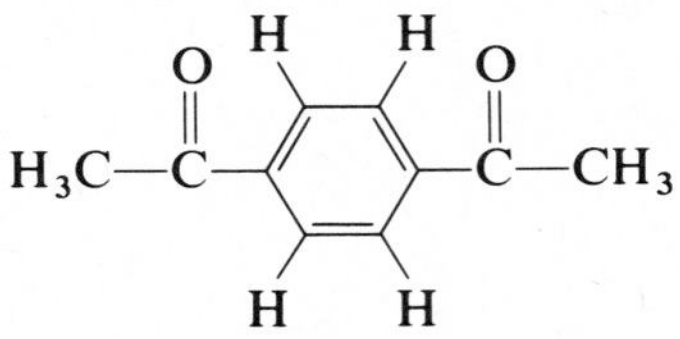

Solution: First, we recognize from symmetry considerations that all four aromatic hydrogens are equivalent. Further, each of them is both ortho *and* meta to a $C{<}^{O}_{R}$ group. Therefore,

$$\delta = 7.27 + 0.63 + 0.27 = 8.17 \text{ ppm}$$

The value actually observed is 8.08 ppm.[2]

EXAMPLE 6-16 Identify the compound C_8H_6 that exhibits signals at δ 7.40 (5H) and 3.09 (1H) in its ^{1}H nmr spectrum.[2]

Solution: Clearly this compound has a monosubstituted aromatic ring, judging from the *five* aromatic hydrogens. About the only possible structure would be

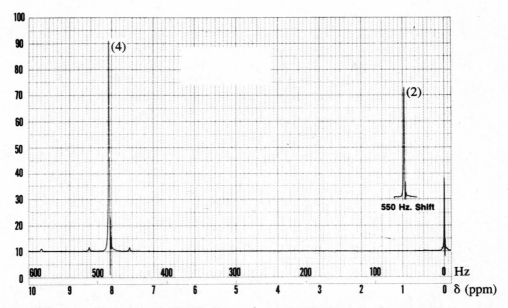

Yet, is the δ 3.09 ppm value a reasonable chemical shift for the acetylenic hydrogen? In Section 6-5 we saw that such hydrogens usually fall in the region δ 2–3 ppm. The observed signal in this case is at the deshielded end of this range, probably because the acetylenic hydrogen lies in the deshielded region around the periphery of the aromatic ring.

EXAMPLE 6-17 Identify the compound ($C_8H_6O_2$) whose 60-MHz ^1H nmr spectrum is shown in Figure 6-6.

FIGURE 6-6 Unknown compound with molecular formula $C_8H_6O_2$ (from *The Aldrich Library of NMR Specta*).

Solution: The ratio of hydrogens and the signal at δ 8.09 ppm suggest a symmetrically disubstituted aromatic ring. But wait a minute! What is this strange peak labeled 550 Hz? Occasionally a peak in the spectrum falls outside the limits of the normal graph. When this happens, the peak can be brought onto the graph by *offsetting* its signal a known amount. In this case the signal was offset 550 Hz, meaning it was moved *to the right* by 550 Hz *after* the rest of the spectrum was plotted. Without this offset, the signal would have been beyond the left edge of the graph. But having offset this signal, how do we calculate its chemical shift? Remember (Section 5-2) that, at 60 MHz, 1 δ unit equals 60 Hz, so 550 Hz is equivalent to 9.17 ppm. Adding this to the apparent position of the signal (0.98 ppm), we calculate an actual shift of 10.15 ppm. But what kind of hydrogens occur at 10 ppm? Have you forgotten already? Look back at Section 6-4. Aldehydic hydrogens! The spectrum belongs to

Note the close similarity of its spectrum to the one described in Example 6-15.

EXAMPLE 6-18 Explain why the two equivalent methyl groups in structure **6-3** appear at δ −4.25 ppm,[4] far *upfield* of TMS.

6-3   side view

Solution: The alternating single and double bonds around the periphery of the fourteen-membered ring give rise to an aromatic electron cloud. The methyl groups lie directly above and below the plane of the ring, in the shielding region of the induced field.

Not all cyclic molecules with alternating single and double bonds are aromatic (see Problem 5 in Self-Test II). A detailed discussion of the molecular orbital theory that predicts whether a molecule is aromatic is beyond the scope of this book. Even if you don't know how to determine when a ring is aromatic, just remember that the shifts of hydrogens attached to the rings in structures such as **6-4** through **6-7** also fall in the aromatic region. Compounds such as **6-4**, **6-6**, and **6-7** are called **heteroaromatic compounds** because their rings contain atoms other than carbon.

pyridine   naphthalene   furan   pyrrole
6-4   **6-5**   **6-6**   **6-7**

6-7. Hydrogen Attached to Elements Other than Carbon

With more than a hundred elements besides carbon in the periodic table (Appendix 1), you might fear that the number of ¹H chemical shift correlations is endless. However, except for a few specialized applications, the most important heteroatoms (those other than carbon) to which hydrogen finds itself attached are oxygen and nitrogen. But before we discuss these two specific cases, here is a useful generalization:

• As the electronegativity of X increases, both the acidity and the chemical shift of a hydrogen attached to X increase.

The electronegativity of elements (except of the inert gases) increases from left to right and from bottom to top in the periodic table, with fluorine being the *most* electronegative.

A. Hydrogen attached to oxygen

A hydrogen attached to an oxygen (O—H) constitutes a **hydroxyl group**. Hydroxyl groups appear in several classes of organic molecules, including **alcohols** (where the carbon bearing the O—H is singly bonded to three other groups), **phenols** (where the O—H is connected directly to an aromatic ring), and **carboxylic acids** (where the O—H group is attached to a carbonyl carbon):

alcohol   phenol   carboxylic acid

For reasons somewhat beyond the scope of this book, the order of acidity of these three classes of compounds is alcohols < phenols < carboxylic acids. Therefore, it shouldn't surprise us that the chemical shifts follow the same order: alcohols (δ 2–4 ppm) < phenols (δ 4.0–7.5 ppm) < carboxylic acids (δ 10–14 ppm). In fact, carboxylic-acid hydrogens are among the most deshielded of all hydrogens, thereby defining the low-field (high-frequency) limit of the ^{1}H spectral range.

However, hydroxyl proton signals have several additional characteristics you should be aware of. To describe these, we need to know a little about the unique properties of acidic hydrogens. Unlike most hydrogens attached to carbon, those attached to oxygen are subject to **hydrogen bonding** and **exchange**. Hydrogen bonding involves the attraction of a relative acidic hydrogen (still bonded to X) by another electronegative atom Y. This is usually written X—H---Y or X---H---Y. With alcohols, this takes the form

$$\text{>}C-O\overset{\displaystyle H}{\underset{\displaystyle H}{\cdots}}O-C\text{<}$$

Under certain conditions (such as the presence of a small amount of acid catalyst), the hydrogens can actually be traded (i.e., exchanged) between the two hydroxyl groups. Such exchange processes can be very rapid, occurring at rates comparable to the nmr time scale (Section 1-4). The results of these two phenomena are that hydroxyl proton signals are often broader (larger half-width, Section 5-1A) than C—H proton signals, and their chemical shifts are quite dependent on temperature, concentration, and the nature of the solvent (Section 3-4C) because all three of these variables affect the rate of hydrogen exchange and the strength of the hydrogen bonds. The upshot of all this is that hydroxyl proton chemical shifts are quite variable, and it is difficult to be as accurate in our predictions of their chemical shifts as we were with hydrogens attached to carbon.

EXAMPLE 6-19 Examine the 60-MHz ^{1}H nmr spectrum of 2,4-pentanediol (Figure 6-7) and decide which signal belongs to the two (equivalent) hydroxyl protons.

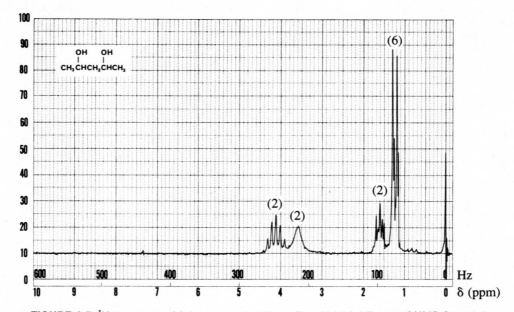

FIGURE 6-7 ^{1}H spectrum of 2,4-pentanediol (from *The Aldrich Library of NMR Spectra*).

Solution: The signal at δ 3.58 ppm is broadened, with a half-width of 12 Hz, so it represents the hydroxyl protons. (You may be wondering why there are so many other signals in this spectrum for what appears to be a symmetrical molecule, but you'll have to wait until Chapter 8 to find out.)

EXAMPLE 6-20 A dilute solution of CH_3OH in $CDCl_3$ (Table 3-1) exhibits its O—H signal at δ 1.43 ppm. When a slight excess of D_2O (deuterated water) is added to the sample, the hydroxyl signal disappears and is replaced by a signal at δ 4.75 ppm.[3] Explain.

Solution: Remember that deuterium (^2H, Section 2-1A) does not show up in a ^1H spectrum. But *chemically* deuterium acts just like hydrogen. There is exchange between the O—H group of the alcohol and the O—D group of D_2O, leading to CH_3OD (which exhibits only a methyl signal) and H—O—D. The latter gives rise to the δ 4.75 ppm signal. This **deuterium exchange** technique is a useful way of confirming the presence of a readily exchangeable hydrogen.

B. Hydrogen attached to nitrogen

Hydrogen bonded to nitrogen occurs in several classes of compounds, as shown in structures **6-8** through **6-11**. Compare these structures with those of alcohols, phenols, and acids. Because nitrogen is somewhat less electronegative than oxygen (why?), N—H proton signals tend to appear at higher field (lower frequency) than their O—H counterparts. The normal chemical shift ranges are amines (δ 0.5–3.0 ppm) < anilines (δ 3.0–5.0 ppm) < amides (δ 4–7 ppm) < ammonium salts (δ 6.0–8.5 ppm). As in the case of O—H hydrogens, N—H signals are also affected by hydrogen bonding and exchange and are therefore also dependent on temperature, concentration, and solvent. There is one further complication. Nuclei with a nuclear spin $\geq$ 1 (Section 2-1C) possess a nonspherical nucleus and, hence, an **electric quadrupole moment**, which hastens the spin–lattice relaxation (Section 2-2B) of nearby nuclei and thereby broadens their signals. Thus protons attached to nitrogen (^{14}N has $I = 1$) can exhibit exceptionally broad signals that are sometimes difficult to differentiate from background noise (Section 3-2C).

amine	aniline	amide	ammonium salt
6-8	**6-9**	**6-10**	**6-11**

EXAMPLE 6-21 Which signal in the 60-MHz ^1H nmr spectrum shown in Figure 6-8 would you assign to the N—H hydrogens in $(CH_3)_3C$—C—NH_2?

(with $\|$ O below the C)

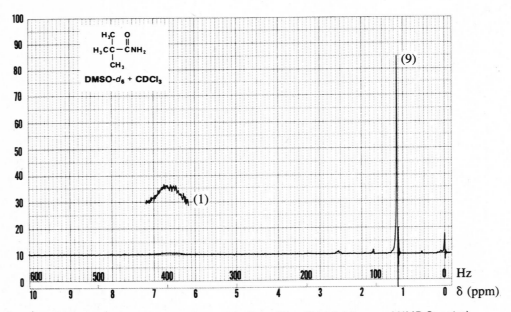

FIGURE 6-8 ^1H spectrum of pivalamide (from *The Aldrich Library of NMR Spectra*).

Solution: The very broad signal centered at 6.6 ppm. It is so broad that in order for it to be seen clearly its intensity must be electronically amplified by increasing the receiver (or plotter) gain. The amplified signal is shown in the upper trace.

In this chapter we have developed a familiarity with the factors that influence the nmr chemical shift of hydrogens, such as their molecular environments. Figure 6-9 reviews graphically what we've learned. We have also begun to use our knowledge by assigning structures to unknown compounds. Let's finish this chapter with one final example to see how well you've mastered the material.

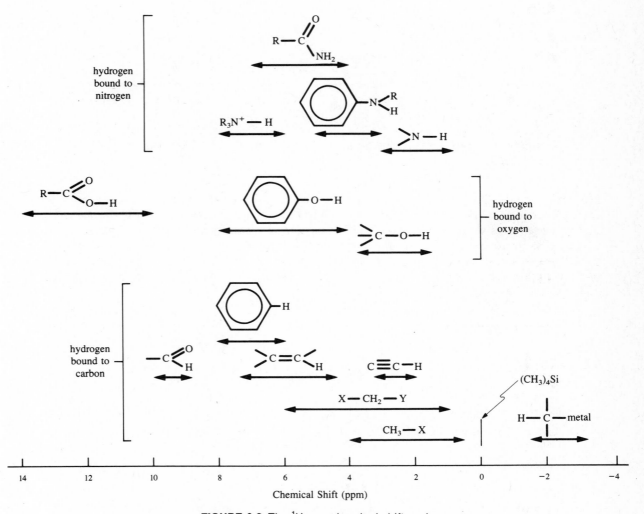

FIGURE 6-9 The ^{1}H nmr chemical shift scale.

EXAMPLE 6-22 Identify the compound $C_2H_8N_2$ whose 60-MHz ^{1}H nmr spectrum is shown in Figure 6-10. Assign both signals to specific hydrogens in your structure and show that their chemical shifts are within expected limits.

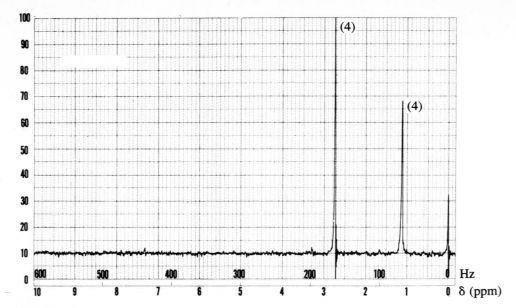

FIGURE 6-10 Unknown compound with molecular formula $C_2H_8N_2$ (from *The Aldrich Library of NMR Spectra*).

Solution: The fact that there are only two signals, each resulting from four hydrogens, suggests a high degree of symmetry in the structure. The signal at δ 1.11 ppm, while equal in intensity to the one at δ 2.73 ppm, is nonetheless shorter and must, therefore, have a somewhat broader half-width (Section 5-1A). This may be due to exchange broadening, suggesting the broader signal is due to hydrogens bonded to nitrogen. Also, a δ 1.11-ppm chemical shift *is* in the range for amine N—H. Moreover, the δ 2.73-ppm signal is approximately correct for R—CH_2—NR_2 [from Eq. (6-6) and Table 6-1 we predict $\delta = 0.23 + 0.47 + 1.57 = 2.27$]. Putting all of this together in one structure leads to

$$H_2N—CH_2—CH_2—NH_2$$

REFERENCES

1. S. Bhacca, D. P. Hollis, L. F. Johnson, and E. A. Pier, *High Resolution NMR Spectra Catalog*. Varian Associates, Palo Alto, 1963.

2. C. J. Pouchert, *The Aldrich Library of NMR Spectra*, 2nd ed. Aldrich Chemical Company, Milwaukee, 1983.

3. R. M. Silverstein, G. C. Bassler, and T. C. Morrill, *Spectrometric Identification of Organic Compounds*, 4th ed. Wiley, New York, 1981.

4. V. Boekelheide and J. B. Phillips, *J. Am. Chem. Soc.* **89**, 1965 (1967).

Other ${}^1H$ nmr spectral compilations, in addition to the above, include:

The Sadtler Standard NMR Spectra. Sadtler Research Laboratories, Philadelphia, 1972.
Handbook of Proton-NMR Spectra and Data. Asahi Research Center, Academic Press, Orlando, 1985.

SUMMARY

1. Although all nuclei of a given isotope have exactly the same magnetogyric ratio, they may precess at slightly different frequencies (and hence give rise to separate nmr signals) because of differences in their molecular environments. In general, the greater the electron density around a nucleus, the more it is shielded from the effects of the applied magnetic field. Therefore, the more shielded a nucleus is, the higher in field strength (or the lower in frequency) will its nmr signal be.

2. Hydrogen nuclei attached to carbon are divided among the following groups:

   —CH$_3$ methyl

   —CH$_2$—   methylene   } aliphatic

   —CH   methine

   =C—H   vinyl (or olefinic)

   ≡C—H   acetylenic

   O=C—H   aldehydic (or formyl)

   ⬡—H   aromatic

3. The chemical shift of a hydrogen attached to carbon can be predicted quite accurately from a version of the following equation, with the appropriate base value and substituent shielding parameter(s):

$$\delta = \text{base value} + \sum \Delta\delta_i$$

4. Certain types of functional groups, most notably carbon–carbon triple bonds and aromatic rings, give rise to anisotropic shielding fields. As a result, acetylenic hydrogens are unusually shielded, while aromatic hydrogens are unusually *de*shielded.

5. Hydrogens attached to oxygen or nitrogen give nmr signals whose positions are highly dependent on concentration, temperature, and solvent. This is because such hydrogens undergo hydrogen bonding and exchange. Often, such signals are quite broad.

7 CHEMICAL SHIFT CORRELATIONS FOR ^{13}C

THIS CHAPTER IS ABOUT

☑ ^{13}C **Chemical Shifts Revisited**
☑ **Singly Bonded Carbons**
☑ **Heterocyclic Structures**
☑ **Vinyl and Aromatic Carbons**
☑ **Triply Bonded Carbons**
☑ **Carbonyl Carbons**
☑ **Summary of** ^{13}C **Chemical Shifts**
☑ **Chemical Shifts of Other Elements**

7-1. ^{13}C Chemical Shifts Revisited

So far, we have developed an understanding of the relationships between the molecular environment of a hydrogen nucleus and the nucleus' chemical shift. Now let's see if the same approach of base values plus shielding parameters will allow us to predict ^{13}C chemical shifts. Recall from Sections 2-1 and 2-2 that ^{13}C (like ^1H) has a nuclear spin of $\frac{1}{2}$ but undergoes resonance at a much lower frequency than ^1H because of its lower magnetogyric ratio. Also, the low natural abundance and low relative sensitivity of ^{13}C (compared to ^1H, Section 2-2) require that the PFT technique (Section 3-3) be used for data collection. And finally, remember that TMS (its *carbon* signal now) still defines the zero point of our chemical shift scale, and that ^{13}C chemical shifts span a range of about 250 ppm.

TABLE 7-1 ^{13}C **Chemical Shifts for Common Alkanes and Cycloalkanes**[a]

Name	Structure	δ (ppm)		
		C_1[b]	C_2	C_3
methane	CH_4	−2.3		
ethane	CH_3CH_3	5.7		
propane	$CH_3CH_2CH_3$	15.8	16.3	
butane	$CH_3CH_2CH_2CH_3$	13.4	25.2	
pentane	$CH_3CH_2CH_2CH_2CH_3$	13.9	22.8	34.7
cyclopropane	$(CH_2)_3$	−3.5		
cyclobutane	$(CH_2)_4$	22.4		
cyclopentane	$(CH_2)_5$	25.6		
cyclohexane	$(CH_2)_6$	26.9		

[a] Data from reference 1 at the end of this chapter.
[b] Numbering starts at a terminal carbon.

7-2. Singly Bonded Carbons

Take a moment to review the ^{13}C spectrum of toluene in Section 5-4. How would you predict the ^{13}C chemical shift of the methyl group in toluene? The logical base value is the ^{13}C chemical shift of methane (which, it turns out, is -2.3 ppm), to which we add the ^{13}C shielding parameter of a phenyl ring connected directly to the methyl carbon ($\Delta\delta = 23$ ppm). The predicted value, therefore, is δ 20.7 ppm, in excellent agreement with the observed value of δ 21.4 ppm. Table 7-1 lists ^{13}C base values for a number of common alkanes and cycloalkanes. Table 7-2 gives the shielding parameters ($\Delta\delta$) for many common substituent groups as a function of their proximity to the carbon (α = one bond separation, β = two bond separations, γ = three bond separations). Notice once again that the deshielding effect of the substituents tends to *decrease* as the number of intervening bonds increases. In fact, most groups exert a modest *shielding* effect (shown by the negative value of $\Delta\delta$) on the γ carbon.

TABLE 7-2 ^{13}C **Substituent Shielding Parameters**[a] $\Delta\delta$ **(ppm)**

X	terminal X $X—C_\alpha—C_\beta—C_\gamma$			internal X $C_\gamma—C_\beta—\overset{\overset{\textstyle X}{\textstyle \mid}}{C_\alpha}—C_\beta—C_\gamma$		
	α	β	γ	α	β	γ
—F	68	9	−4	63	6	−4
—NO$_2$	63	4		57	4	
—OR	58	8	−4	51	5	−4
—O—C(=O)R	51	6	−3	45	5	−3
—OH	48	10	−5	41	8	−5
—NR$_2$	42	6	−3			−3
—Cl	31	11	−4	32	10	−4
—C(=O)R	30	1	−2	24	1	−2
—NH$_2$	29	11	−5	24	10	−5
—phenyl	23	9	−2	17	7	−2
—C(=O)NH$_2$	22		−0.5	2.5		−0.5
—C(=O)OH	21	3	−2	16	2	−2
—CH=CH$_2$	20	6	−0.5			−0.5
—C(=O)OR	20	3	−2	17	2	−2
—Br	20	11	−3	25	10	−3
—CH$_3$	9	10	−2	6	8	−2
—C≡C—H	4.5	5.5	−3.5			−3.5
—C≡N	4	3	−3	1	3	−3
—I	−6	11	−1	4	12	−1

[a] Data from Wehrli, F. W., and Wirthlin, T., *Interpretation of Carbon-13 NMR Spectra*, Heyden, London, 1976, as quoted in reference 1 at the end of this chapter.

EXAMPLE 7-1 Why does butane (Table 7-1) exhibit only two ^{13}C signals, rather than four?

Solution: By symmetry, the two methyl groups are equivalent (δ 13.4 ppm), as are the two methylenes (δ 25.2 ppm).

EXAMPLE 7-2 Predict the chemical shift of each carbon in

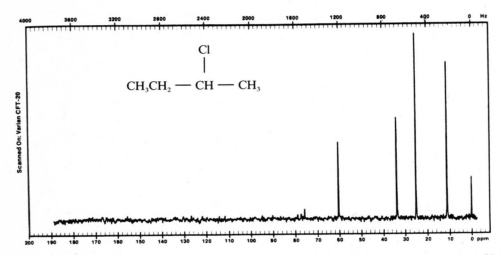

Solution: Using as a base value the chemical shift of cyclopentane (δ 25.6 ppm, Table 7-1) and the internal shielding parameters of the —OH group, we calculate:

$$\delta_\alpha = \delta(\text{cyclopentane}) + \Delta\delta(\text{OH}, \alpha) = 25.6 + 41 = 66.6 \text{ ppm}$$

$$\delta_\beta = \delta(\text{cyclopentane}) + \Delta\delta(\text{OH}, \beta) = 25.6 + 8 = 33.6 \text{ ppm}$$

$$\delta_\gamma = \delta(\text{cyclopentane}) + \Delta\delta(\text{OH}, \gamma) = 25.6 + (-5) = 20.6 \text{ ppm}$$

These are in reasonable agreement with the observed values of δ 73.3, 35.0, and 23.4 ppm.[1]

EXAMPLE 7-3 The ^{13}C spectrum of 2-chlorobutane is reproduced in Figure 7-1. (*note:* All ^{13}C spectra reproduced in this chapter are proton-decoupled. The importance of this will be described in Chapter 11.) Assign each signal to a carbon in the structure by correlating the observed chemical shifts with the predicted values.

FIGURE 7-1 The 20-MHz ^{13}C spectrum of 2-chlorobutane. © Sadtler Research Laboratories, Division of Bio-Rad Laboratories, Inc., 1983.

Solution:

For carbon 1 we use the butane base value of δ 13.4 ppm (Table 7-1) and add the $\Delta\delta$ value for an internal β chlorine (10 ppm, Table 7-2). Thus,

$$\delta_1 = \delta(\text{butane}, \text{C}_1) + \Delta\delta(\text{Cl}, \beta) = 13.4 + 10 = 23.4 \text{ ppm}$$

Similarly,

$$\delta_2 = \delta(\text{butane}, \text{C}_2) + \Delta\delta(\text{Cl}, \alpha) = 25.2 + 32 = 57.2 \text{ ppm}$$

$$\delta_3 = \delta(\text{butane}, \text{C}_2) + \Delta\delta(\text{Cl}, \beta) = 25.2 + 10 = 35.2 \text{ ppm}$$

$$\delta_4 = \delta(\text{butane}, \text{C}_1) + \Delta\delta(\text{Cl}, \gamma) = 13.4 + (-4) = 9.4 \text{ ppm}$$

Clearly, δ_1 corresponds to the peak at δ 25.0, δ_2 to the peak at δ 60.1, δ_3 to the peak at δ 33.6, and δ_4 to the peak at δ 11.1.

Note how well the predicted values compare with the actual values. (^{13}C chemical shifts can be ascertained only to ± 1 ppm when measured directly off a plotted spectrum. These more precise values come from a computer listing that accompanies the spectrum.)

By the way, did you notice the relative intensities of the lines in Figure 7-1? From Section 5-4A we know that ^{13}C signal intensities are not necessarily proportional to numbers of carbons (as they are in the case of hydrogen signals) because of differences in relaxation times and NOE enhancements. Nonetheless, as is usually the case, the relative intensity of the four carbon signals in the spectrum of 2-chlorobutane varies according to the number of hydrogens directly attached to the carbon, i.e., methyl > methylene > methine.

EXAMPLE 7-4 Okay, now let's try to identify an unknown. Suggest a structure for $C_5H_{13}N$, whose ^{13}C spectrum is shown in Figure 7-2. Assign each signal to a carbon in your structure, then calculate the expected chemical shift of each carbon.

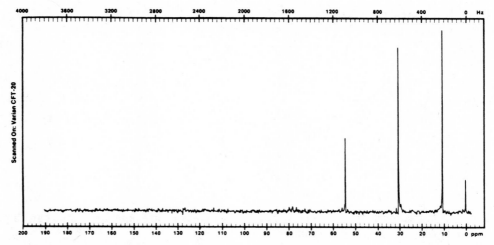

FIGURE 7-2 The 20-MHz ^{13}C spectrum of unknown compound with molecular formula $C_5H_{13}N$. © Sadtler Research Laboratories, Division of Bio-Rad Laboratories, Inc., 1983.

Solution: The fact that there are only three signals (δ 54.4, 30.4, and 10.4 ppm) for the five carbons suggests that the structure has some symmetry. The only substituent groups with a single nitrogen that

$$R-NH_2 \qquad \begin{matrix} R \\ R \end{matrix}\!\!\Big\rangle N-H \qquad \begin{matrix} R \\ R \end{matrix}\!\!\Big\rangle N-R$$

1° amine   2° amine   3° amine
7-1   **7-2**   **7-3**

give a signal in the δ 50–60 ppm region are primary (1°), secondary (2°), or tertiary (3°) amines (structures **7-1** through **7-3**). The signal at δ 54.4 ppm represents either one carbon or two or three *equivalent* carbons attached to nitrogen. A little thought should convince you that there is no way to distribute *five* carbons among two or three equivalent R groups to yield a secondary or tertiary amine unless there were *two* different signals in the δ 50–60 ppm range. Therefore, we must be dealing with a *primary* amine having just *one* carbon attached to nitrogen. The four remaining carbons must be divided into two sets of two to account for the remaining signals. The only structure that fulfills these requirements is

$$
\begin{array}{l}
\overset{1}{H_3C}-\overset{2}{CH_2} \\
\qquad\qquad\overset{3}{H-C-NH_2} \\
H_3C-CH_2
\end{array}
$$

To confirm our assignment, let's calculate the predicted chemical shifts, using the base values for pentane (Table 7-1):

$$\delta_1 = \delta(\text{pentane, } C_1) + \Delta\delta(NH_2, \gamma) = 13.9 + (-5) = 8.9 \text{ ppm}$$

$$\delta_2 = \delta(\text{pentane, } C_2) + \Delta\delta(NH_2, \beta) = 22.8 + 10 = 32.8 \text{ ppm}$$

$$\delta_3 = \delta(\text{pentane, } C_3) + \Delta\delta(NH_2, \alpha) = 34.7 + 24 = 58.7 \text{ ppm}$$

Once again, the agreement is close enough to give us confidence in our structural assignment.

Did you notice the three very small peaks (barely discernable from background noise) in the 75–80-ppm region of the spectra in Figures 7-1 and 7-2? Do you remember what causes these? If not, take a look back at Section 5-4A.

7-3. Heterocyclic Structures

A cyclic molecule in which one or more of the ring atoms is a *heteroatom* (an atom other than carbon) is referred to as a **heterocyclic molecule**.

EXAMPLE 7-5 Predict the chemical shift of the carbons in this heterocyclic molecule:

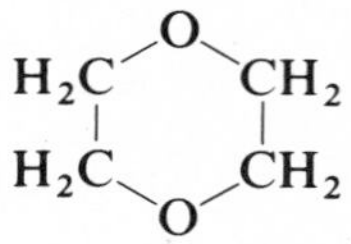

Solution: First, we recognize by symmetry that all four carbons are equivalent. We can think of the molecule as if it were a derivative of ethane:

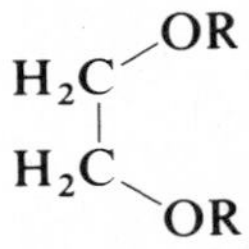

Thus, each carbon is α to one and β to another (equivalent) O—R group. Using the data in Tables 7-1 and 7-2, we calculate

$$\delta = \delta(\text{ethane}) + \Delta\delta(OR, \alpha) + \Delta\delta(OR, \beta) = 5.7 + 58 + 8 = 71.7 \text{ ppm}$$

The observed value is 67.4 ppm.[1]

When predicting chemical shifts for substituted heterocycles, you will find it preferable to use the chemical shifts of the parent heterocycle as base values. A few typical examples are shown in structures **7-4**, **7-5**, and **7-6**[1,4]. Consult the references listed at the end of this chapter for additional examples.

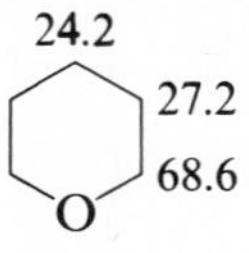

tetrahydropyran	piperidine	thiacyclohexane
7-4	**7-5**	**7-6**

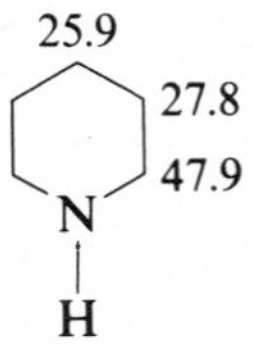

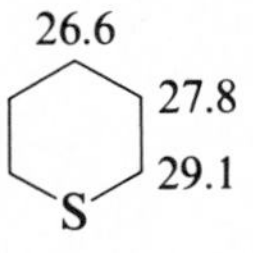

EXAMPLE 7-6 Predict the chemical shift of each carbon in the following structure:

$$\begin{array}{c} 1 \rightarrow CH_3 \\ 4 \rightarrow \\ 3 \rightarrow \\ 2 \rightarrow \quad N \\ \quad H \end{array}$$

Solution: Using the base values in structure **7-5** and the shielding parameters for an internal CH_3 group (Table 7-2) we predict for the ring carbons:

$$\delta_2 = \delta(\text{piperidine}, C_2) + \Delta\delta(CH_3, \gamma) = 47.9 + (-2) = 45.9 \text{ ppm}$$

$$\delta_3 = \delta(\text{piperidine}, C_3) + \Delta\delta(CH_3, \beta) = 27.8 + 8 = 35.8 \text{ ppm}$$

$$\delta_4 = \delta(\text{piperidine}, C_4) + \Delta\delta(CH_3, \alpha) = 25.9 + 6 = 31.9 \text{ ppm}$$

We can calculate the chemical shift of the methyl carbon by viewing the molecule as a substituted derivative of ethane (Table 7-2):

$$\begin{array}{c} R \\ | \\ H_3C-C-H \\ | \\ R \end{array}$$

Thus

$$\delta_4 = \delta(\text{ethane}) + \Delta\delta(CH_3, \beta) + \Delta\delta(CH_3, \beta) = 5.7 + 10 + 10 = 25.7 \text{ ppm}$$

These turn out to be very close to the actual values of δ 46.8, 35.7, 31.3, and 22.5 ppm, respectively.[3]

7-4. Vinyl and Aromatic Carbons

You probably remember that the *aromatic* (ring) carbons of toluene (Section 5-4) appear far downfield of the methyl carbon. Just as with vinyl and aromatic *hydrogens* (Sections 6-4 and 6-6), vinyl and aromatic carbons are quite deshielded in comparison to their singly bonded counterparts. For this reason, their signals usually fall in the region of δ 100–160 ppm.

We'll treat molecules with vinyl carbons as derivatives of ethylene (Section 6-4), whose ^{13}C signal appears at δ 123.3 ppm.[1] To this base value, we apply $\Delta\delta$ corrections (Table 7-3) for substituents according to their location with respect to the carbon.

EXAMPLE 7-7 Predict the chemical shift of the two vinyl carbons in

$$\begin{array}{c} H \qquad\qquad CH_3 \\ \diagdown \quad C=C \quad \diagup \\ CH_3-CH_2 \quad \uparrow \quad \uparrow \quad H \\ \qquad\qquad 1 \quad 2 \end{array}$$

Solution: Carbon 1 has alkyl carbon substituents at positions α, β, and α'. Therefore,

$$\delta_1 = \delta(\text{ethylene}) + \Delta\delta(C, \alpha) + \Delta\delta(C, \beta) + \Delta\delta(C, \alpha') = 123.3 + 10.6 + 7.2 + (-7.9) = 133.2 \text{ ppm}$$

Similarly,

$$\delta_2 = \delta(\text{ethylene}) + \Delta\delta(C, \alpha) + \Delta\delta(C, \alpha') + \Delta\delta(C, \beta') = 123.3 + 10.6 + (-7.9) + (-1.8) = 124.2 \text{ ppm}$$

The observed values are ... (may I have the envelope please?) δ 133.3 and 123.7 ppm, respectively.[1]

TABLE 7-3 Vinyl Carbon Substituent Shielding Parameters[a]

$$X_\gamma\!-\!X_\beta\!-\!X_\alpha\!\diagdown \underset{X_\alpha}{\overset{}{C}}\!=\!\underset{X_{\alpha'}}{\overset{}{C}}\diagup^{X_{\alpha'}\!-\!X_{\beta'}\!-\!X_{\gamma'}}$$

this carbon

	$\Delta\delta$ (ppm)					
X	α	β^b	γ^b	α'	β'^b	γ'^b
—C≡	10.6	7.2	−1.5	−7.9	−1.8	−1.5
—OR	29	2		−39	−1	
—O—C(=O)R	18			−27		
—C(=O)R	15			6		
—phenyl	12			−11		
—C(=O)OR	6			7		
—C(=O)OH	4			9		
—Cl	3	−1		−6	2	
—Br	−8	0		−1	2	
—C≡N	−16			15		
—I	−38			7		

[a] Data from references 1 and 4 at the end of this chapter.
[b] When a group is in the β position, X_α is assumed to be carbon; when a group is in the γ position, both X_α and X_β are assumed to be carbon.

Table 7-3 includes shielding parameters for several heteroatom-containing substituents. Notice that there is a very large difference between the $\Delta\delta$ value for a group located at the α position and the value for the same group at the α' position. This is because the electrons in the carbon–carbon double bond interact directly with many of these substituents, causing substantial shielding when at the α position and deshielding when at the α' position, or vice versa. This interaction is called **resonance** or **conjugation**.

EXAMPLE 7-8 Predict the vinyl-carbon chemical shifts for the molecule

$$H_3C\!-\!O\diagdown \underset{H\uparrow}{\overset{}{C}}\!=\!\underset{\underset{2}{\uparrow}H}{\overset{}{C}}\diagup^{H}$$

$$\underset{1}{}\quad\underset{2}{}$$

Solution: Using the data in Table 7-3,

$$\delta_1 = \delta(\text{ethylene}) + \Delta\delta(\text{OR}, \alpha) = 123.3 + 29 = 152.3 \text{ ppm}$$

$$\delta_2 = \delta(\text{ethylene}) + \Delta\delta(\text{OR}, \alpha') = 123.3 + (-39) = 84.3 \text{ ppm}$$

The observed signals occur at δ 153.2 and 84.2 ppm, respectively.[1] Because of resonance interactions, the best policy to follow when you are predicting chemical shifts of vinyl carbons with such *polar* substituents is to find a model compound as similar in structure as possible to serve as the source of base values.

TABLE 7-4 Aromatic Substituent Shielding Parameters[a]

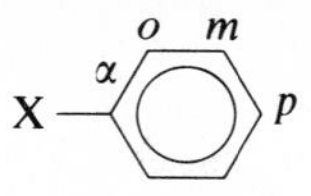

$$X \text{—} \overset{\alpha}{\underset{}{\bigcirc}} \overset{o \quad m}{\underset{}{}} p$$

$\Delta\delta$ (ppm)

X	α	o(ortho)	m(meta)	p(para)
—F	34.8	−12.9	1.4	−4.5
—OCH$_3$	31.4	−14.4	1.0	−7.7
—OH	26.9	−12.7	1.4	−7.3
—O—C(=O)R	23.0	−6.4	1.3	−2.3
—NR$_2$	22.4	−15.7	0.8	−15.7
—NO$_2$	20.0	−4.8	0.9	5.8
—NH$_2$	18.0	−13.3	0.9	−9.8
—CH$_2$CH$_3$	15.6	−0.5	0.0	−2.6
—phenyl	13.1	−1.1	0.4	−1.2
—CH=CH$_2$	9.5	−2.0	0.2	−0.5
—C(=O)R	9.1	0.1	0.0	4.2
—CH$_3$	8.9	0.7	−0.1	−2.9
—C(=O)H	8.6	1.3	0.6	5.5
—Cl	6.2	0.4	1.3	−1.9
—C(=O)OH	2.1	1.5	0.0	5.1
—C(=O)OR	1.3	−0.5	−0.5	3.5
—Br	−5.5	3.4	1.7	−1.6
—C≡CH	−6.1	3.8	0.4	−0.2
—C≡N	−15.4	3.6	0.6	3.9
—I	−32.2	9.9	2.6	−7.4

[a] Data from reference 1 at the end of this chapter.

In the case of aromatic carbons, we will use as our base value the ^{13}C chemical shift of benzene (Example 4-3), δ 128.5 ppm. Notice that aromatic *carbons* appear slightly downfield of vinyl carbons, just as aromatic *hydrogens* appear downfield of vinyl hydrogens (Sections 6-4 and 6-6). Table 7-4 lists substituent shielding parameters for a number of aromatic substituents as a function of their location. Observe how, for many of these groups, their effect alternates back and forth from deshielding to shielding as the number of intervening bonds increases.

EXAMPLE 7-9 Predict the chemical shifts of each aromatic carbon in

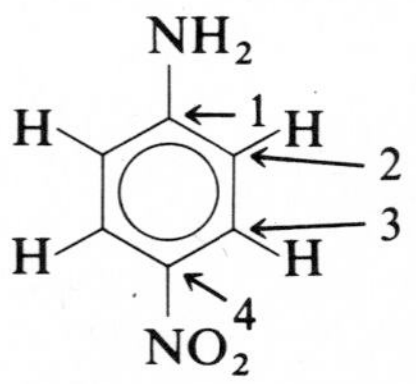

Solution: By symmetry, the aromatic carbons are divided into four sets in the ratio of 1:2:2:1 carbons. Carbon 1 is α to an NH_2 group, and para to a NO_2 group. Therefore,

$$\delta_1 = \delta(benzene) + \Delta\delta(NH_2, \alpha) + \Delta\delta(NO_2, p) = 128.5 + 18.0 + 5.8 = 152.3 \text{ ppm}$$

Likewise,

$$\delta_2 = \delta(benzene) + \Delta\delta(NH_2, o) + \Delta\delta(NO_2, m) = 128.5 + (-13.3) + 0.9 = 116.1 \text{ ppm}$$

$$\delta_3 = \delta(benzene) + \Delta\delta(NH_2, m) + \Delta\delta(NO_2, o) = 128.5 + 0.9 + (-4.8) = 124.6 \text{ ppm}$$

$$\delta_4 = \delta(benzene) + \Delta\delta(NH_2, p) + \Delta\delta(NO_2, \alpha) = 128.5 + (-9.8) + 20.0 = 138.7 \text{ ppm}$$

These predictions agree quite nicely with the observed values of δ 155.1, 112.8, 126.3, and 136.9 ppm, respectively.[2]

EXAMPLE 7-10 Which of the three isomers of xylene gives the ^{13}C spectrum shown in Figure 7-3? To confirm your choice, calculate the expected chemical shift of each carbon in your structure.

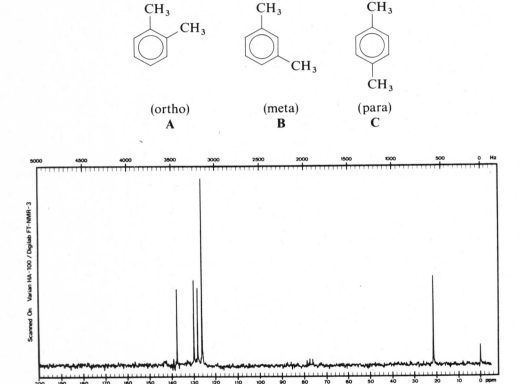

FIGURE 7-3 The 25-MHz ^{13}C spectrum of an isomer of xylene. ©Sadtler Research Laboratories, Division of Bio-Rad Laboratories, Inc., 1983.

Solution: If you took more than one minute to pick out the correct answer, you've forgotten all you've learned about symmetry. Go back three chapters and review the topic! In each of these three structures the two methyl groups are equivalent. But the number of aromatic carbon signals would vary: **A** would show *three*, **B** would show *four*, while **C** would show only *two*. Thus, only **B** fits the observed spectrum.

To confirm our diagnosis, let's predict chemical shifts.

$$\delta_1 = \delta(\text{benzene}) + \Delta\delta(CH_3, o) + \Delta\delta(CH_3, o) = 128.5 + 0.7 + 0.7 = 129.9 \text{ ppm}$$

$$\delta_2 = \delta(\text{benzene}) + \Delta\delta(CH_3, \alpha) + \Delta\delta(CH_3, m) = 128.5 + 8.9 + (-0.1) = 137.3 \text{ ppm}$$

$$\delta_3 = \delta(\text{benzene}) + \Delta\delta(CH_3, o) + \Delta\delta(CH_3, p) = 128.5 + 0.7 + (-2.9) = 126.3 \text{ ppm}$$

$$\delta_4 = \delta(\text{benzene}) + \Delta\delta(CH_3, m) + \Delta\delta(CH_3, m) = 128.5 + (-0.1) + (-0.1) = 128.3 \text{ ppm}$$

The observed values from the spectrum are δ 130.0, 137.6, 126.2, and 128.2 ppm, respectively. The observed chemical shift of the methyl groups (δ 21.3 ppm) could have been predicted from the data for toluene in Section 7-2.

As was true for the *hydrogens* attached to heteroaromatic rings (Section 6-6), the *carbons* of such rings also appear in the aromatic region. Some examples[1] are

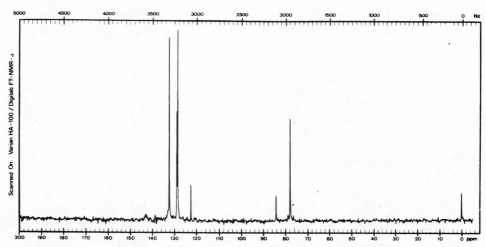

pyridine   pyrrole   furan

7-5. Triply Bonded Carbons

Do you recall (Section 6-5) that hydrogens attached to triply bonded carbons (*acetylenic* hydrogens) are unusually *shielded*? Exactly the same is true of acetylenic carbons themselves. Triply bonded carbons usually appear in the δ 70–90-ppm region, significantly *upfield* of typical vinyl and aromatic carbons. This a region of the spectrum where few other types of carbons are found. Acetylenic carbons are classified as either *terminal* (if they have a hydrogen directly attached), or *internal* (if they have a carbon attached). Signals for internal acetylenic carbons are usually downfield of, and less intense than, the signals for terminal ones. This is because of the deshielding effect of the carbon substituent and the absence of the intensity-increasing NOE (Section 5-4A) of an attached hydrogen.

EXAMPLE 7-11 Propose a structure for the compound C_8H_6 whose ^{13}C spectrum is shown in Figure 7-4. Assign all signals by calculating expected chemical shifts.

FIGURE 7-4 The 25-MHz ^{13}C spectrum of unknown compound with molecular formula C_8H_6. © Sadtler Research Laboratories, Division of Bio-Rad Laboratories, Inc., 1983.

Solution: We note immediately that there are four signals in the aromatic region and two in the acetylenic region. Thus, a likely candidate would be an aromatic ring with one acetylenic substituent:

$$\underset{6}{\overset{5\quad 4}{\bigcirc}}\overset{3}{-}\overset{2}{C}\equiv\overset{1}{C}H$$

Carbons 1 and 2 can be assigned on the basis of intensity (remember the NOE: fewer attached hydrogens, weaker signal) and relative position to the signals at δ 77.4 and 83.8 ppm, respectively. To assign the aromatic signals, let's compare predicted with observed values.

Carbon	$\delta_{predicted}$	$\delta_{observed}$
3	$\delta(\text{benzene}) + \Delta\delta(-C_2H, \alpha) = 128.5 + (-6.1) = 122.4$ ppm	122.4 ppm
4	$\delta(\text{benzene}) + \Delta\delta(-C_2H, o) = 128.5 + 3.8 = 132.3$ ppm	132.2 ppm
5	$\delta(\text{benzene}) + \Delta\delta(-C_2H, m) = 128.5 + 0.4 = 128.9$ ppm	128.3 ppm
6	$\delta(\text{benzene}) + \Delta\delta(-C_2H, p) = 128.5 + (-0.2) = 128.3$ ppm	128.7 ppm

Note that the signals for carbons 5 (of which there are two) and 6 (of which there is one) are so close that their assignments are based on relative signal intensity rather than on the small difference in their δ values.

Incidentally, does this structure remind you of anything? It should! Look back at Example 6-16. Notice once again how the relative chemical shifts of the *hydrogens* parallel the relative chemical shifts of the *carbons* to which they're attached.

Another important triply bonded carbon is the one in a $-C\equiv N$ group (**cyano group**). Because of the higher electronegativity (and hence deshielding effect) of the nitrogen, cyano carbons occur somewhat downfield of acetylenic carbons, usually around δ 115–120 ppm. And remember: With no hydrogens attached, they usually give fairly weak signals.

TABLE 7-5 ^{13}C Chemical Shift Ranges of Carbonyl Compounds

Compound class	Structure	δ (ppm)
ketone	$R-\overset{\overset{\displaystyle O}{\|}}{C}-R$	195–220
aldehyde	$R-\overset{\overset{\displaystyle O}{\|}}{C}-H$	190–200
carboxylic acid	$R-\overset{\overset{\displaystyle O}{\|}}{C}-OH$	170–185
carboxylate ester	$R-\overset{\overset{\displaystyle O}{\|}}{C}-OR$	165–175
anhydride	$R-\overset{\overset{\displaystyle O}{\|}}{C}-O-\overset{\overset{\displaystyle O}{\|}}{C}-R$	165–175
amide	$R-\overset{\overset{\displaystyle O}{\|}}{C}-NR_2$	160–170
acid halide	$R-\overset{\overset{\displaystyle O}{\|}}{C}-X$ (X = Cl, Br, I)	160–170

7-6. Carbonyl Carbons

The most deshielded carbon atoms are those that are doubly bonded to oxygen (C=O). Because of the high electronegativity of oxygen, carbonyl carbons generally appear in the δ 165–220-ppm range. The carbonyl group occurs in many types of compounds, and its ^{13}C chemical shift varies accordingly, as shown in Table 7-5. Because the regions characteristic of each functional group tend to overlap, one must often consider the chemical shifts of the other carbons in order to suggest a unique structure for an unknown compound.

In addition to their characteristic downfield positions, carbonyl carbon signals can also be recognized by their generally low intensities. Except in the cases of aldehydes and formates (H—C$\diagup^{\displaystyle O}_{\displaystyle O—R}$),

there are no hydrogens directly attached to the carbonyl carbon, so their signals cannot benefit from NOE enhancement (Section 5-4A).

EXAMPLE 7-12 To which of the isomeric structures below does the ^{13}C spectrum in Figure 7-5 belong? Assign all signals by correlating them with predicted chemical shifts.

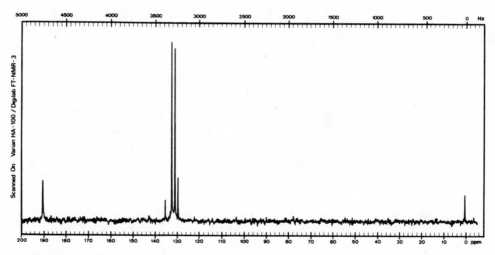

FIGURE 7-5 The 25-MHz ^{13}C spectrum of unknown compound with molecular formula C_7H_5BrO. © Sadtler Research Laboratories, Division of Bio-Rad Laboratories, Inc., 1983.

Solution: It's easy to pick out the carbonyl signal; it's the one at δ 190.6 ppm. From Table 7-5 we can see that this signal is within the normal range for aldehydic carbons, but too far downfield for an acid halide carbonyl. So, we pick the structure on the right. To confirm our selection, let's predict the chemical shifts of the aromatic carbons.

Carbon	$\delta_{\text{predicted}}$	δ_{observed}
1	$\delta(\text{benzene}) + \Delta\delta(\text{—CHO}, \alpha) + \Delta\delta(\text{Br}, p) = 128.5 + 8.6 + (-1.6) = 135.5$ ppm	135.2 ppm
2	$\delta(\text{benzene}) + \Delta\delta(\text{—CHO}, o) + \Delta\delta(\text{Br}, m) = 128.5 + 1.3 + 1.7 = 131.5$ ppm	130.8 ppm
3	$\delta(\text{benzene}) + \Delta\delta(\text{—CHO}, m) + \Delta\delta(\text{Br}, o) = 128.5 + 0.6 + 3.4 = 132.5$ ppm	132.3 ppm
4	$\delta(\text{benzene}) + \Delta\delta(\text{—CHO}, p) + \Delta\delta(\text{Br}, \alpha) = 128.5 + 5.5 + (-5.5) = 128.5$ ppm	129.4 ppm

Q.E.D.!

There is one other type of carbon besides carbonyl that appears in the same downfield region of the ^{13}C spectrum. It is the *central* carbon of an **allene** linkage (C=C=C). Such carbons give very weak signals (why?) in the δ 200–215-ppm region.[1] The two outer carbons, by contrast, are more *shielded* than typical vinyl carbons (Section 7-4) and appear in the δ 75–95-ppm range.[1]

7-7. Summary of ^{13}C Chemical Shifts

The range of ^{13}C chemical shifts is shown graphically in Figure 7-6. This is a useful place to start when trying to identify an unknown compound from its ^{13}C spectrum. But remember that these ranges represent only generalizations and that certain combinations of substituents will cause a signal to show up outside the confines of its "normal" region. Also, notice the similarity of this figure with the corresponding one for *hydrogen* chemical shifts (Figure 6-9). Do you see how parallel the two are? This is because *in general* (there *are* exceptions)

- A carbon and a hydrogen attached directly to it experience similar shielding and deshielding effects of neighboring substituents.

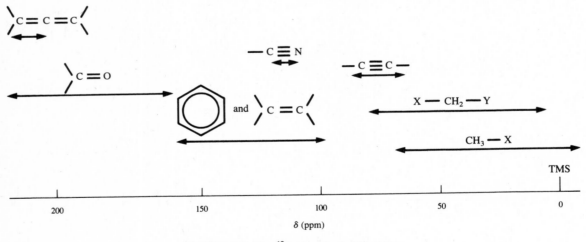

FIGURE 7-6 The ^{13}C chemical shift range.

EXAMPLE 7-13 Suggest a structure for C_8H_5NO, whose ^{13}C spectrum is shown in Figure 7-7. Assign all signals by calculating the expected chemical shift of each carbon in your suggested structure. (*note:* The small peak at δ 128.5 ppm, marked with an "x," is due to an impurity. Can you guess what that impurity might be?)

Solution: There are six signals in this spectrum. From Figure 7-6 we can make some *tentative* guesses. The four signals at δ 129.6, 130.3, 133.4, and 136.9 ppm fall in the middle of the aromatic carbon region, while the one at δ 167.8 ppm is likely a carbonyl carbon. But what nitrogen-containing group would appear at δ 112.9 ppm? Aha! A cyano group! Now, see if you can put all this into some tentative

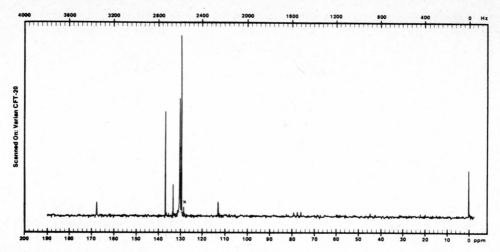

FIGURE 7-7 The 20-MHz ^{13}C spectrum of unknown compound with molecular formula C_8H_5NO. © Sadtler Research Laboratories, Division of Bio-Rad Laboratories, Inc., 1983.

structures:

We can eliminate the first two of these on the basis of symmetry (how many carbon signals would each exhibit?) Come to think of it, the third structure isn't too good either. Its (aldehydic) carbonyl should appear around δ 190 ppm, not 168 ppm. Can you think of any other possibilities? How about

Now, admittedly, this is a ketone of sorts. So we might have expected its carbonyl to appear around δ 200 ppm. But shielding by the triple bond of the directly attached cyano group (Section 6-5) might cause the observed upfield shift. Let's see how close the aromatic carbons come to our expectations for such a structure, using the $\Delta\delta$ for $-C\overset{\displaystyle O}{\underset{\displaystyle R}{\big\|}}$ from Table 7-4:

$$\delta_1 = \delta(\text{benzene}) + \Delta\delta(-C\overset{O}{\underset{R}{}}, \alpha) = 128.5 + 9.1 = 137.6 \text{ ppm}$$

$$\delta_2 = \delta(\text{benzene}) + \Delta\delta(-C\overset{O}{\underset{R}{}}, o) = 128.5 + 0.1 = 128.6 \text{ ppm}$$

$$\delta_3 = \delta(\text{benzene}) + \Delta\delta(-C\overset{O}{\underset{R}{}}, m) = 128.5 + 0.0 = 128.5 \text{ ppm}$$

$$\delta_4 = \delta(\text{benzene}) + \Delta\delta(-C\overset{O}{\underset{R}{}}, p) = 128.5 + 4.2 = 132.7 \text{ ppm}$$

Still, we had best exercise a little caution before assigning the actual signals. In the structure above, which *aromatic* carbon signal should have the lowest intensity? Right! Carbon 1, since it has no hydrogens attached. Carbon 4 (of which there is only one) should be next in intensity. Because there are two carbon 2's and two carbon 3's, their signals should be somewhat more intense. Thus, based on intensities, our assignments would be:

Carbon	$\delta_{observed}$
1	133.4 ppm
2	130.3 ppm
3	129.6 ppm
4	136.9 ppm

Notice that the *relative* positions of carbons 1 and 4 are reversed from the calculated order. So, which do we trust more, calculated chemical shifts or relative intensities? The best advice is to find a similar compound in the literature with which to compare. If such data are not available, then depend on *calculations* if the substituents can be *confidently* assumed to be truly independent, or on *intensities* if they cannot. Just to set your mind at ease, the cyano–ketone structure above *is* the correct answer. This example demonstrates that in many real-life situations we have to use all the data at our disposal and still make some educated judgments. That, after all, is what makes life interesting!

Oh, I almost forgot. The impurity at δ 128.5 ppm could very likely be benzene!

7-8. Chemical Shifts of Other Elements

As suggested in Chapters 2 and 5, the applications of nmr extend well beyond just hydrogen and carbon, although these two are arguably the most important. Other nuclei subjected to *routine* nmr examination include ^2H, ^{15}N, ^{19}F, and ^{31}P; many others are studied nonroutinely. For many of these other nuclei, semi-empirical chemical shift correlations of the type we've seen for ^1H and ^{13}C have been generated. If you're interested in these applications, please refer to more detailed texts such as reference 5.

REFERENCES

1. R. M. Silverstein, G. C. Bassler, and T. C. Morrill, *Spectrometric Identification of Organic Compounds*, 4th ed. Wiley, New York, 1981.
2. W. W. Simons (ed.), *The Sadtler Guide to Carbon-13 NMR Spectra*. Sadtler Research Laboratories, Division of Bio-Rad Laboratories, Inc., Philadelphia, 1983.
3. L. F. Johnson and W. C. Janlowski, *Carbon-13 NMR Spectra*. Krieger, Huntington, N.Y., 1978.
4. J. W. Cooper, *Spectroscopic Techniques for Organic Chemists*. Wiley, New York, 1980.
5. E. D. Becker, *High Resolution NMR*, 2nd ed. Academic Press, New York, 1980.

SUMMARY

1. Carbon atoms in molecules are classified, as are hydrogens, on the basis of the structural unit of which they are a part:

$-CH_3$ methyl $-CH_2-$ methylene $-\overset{|}{C}H$ methine

$-\overset{|}{\underset{|}{C}}-$ quaternary $>C=C<$ vinyl $-C\equiv C-$ acetylenic

aromatic $>C=O$ carbonyl

2. Because of the low natural abundance and low sensitivity of ¹³C, the acquisition of carbon nmr spectra requires the use of pulsed Fourier transform (PFT) methods.

3. The chemical shift of each carbon in a molecule can be predicted accurately by the same type of calculations used for hydrogens:

$$\delta_C = \text{base value} + \sum \Delta\delta$$

where the base value is selected for the appropriate class of carbon, and $\Delta\delta$ represents the substitutent shielding parameters.

4. Generally, the relative values of carbon chemical shifts parallel the relative values of the chemical shifts of the hydrogens attached to the carbons, though the span of ¹³C chemical shifts is much greater than the span of ¹H chemical shifts.

SELF-TEST I

At this point it's time to demonstrate that you can really apply the things you've learned so far in this book. If you can solve the four problems below *without looking at the answers in Appendix 2*, you're ready to proceed to the next topics. If not, perhaps you should review the relevant sections of the first seven chapters. Also, if your background in chemistry is weak, remember that the usual number of bonds to (the valence of) the common elements are C: 4; H: 1; O: 2; N: 3. Refer back as necessary to the previous chapters to get values for physical constants, $\Delta\delta$'s, etc.

1. Consider the ^1H nmr spectrum shown in Figure I-1.

   (a) In this spectrum, 0 to 8.0 ppm on the δ scale corresponds to 480 Hz. At what nominal frequency was this spectrum obtained? In what part of the electromagnetic spectrum does radiation of this frequency appear? What is the energy of a photon of this frequency?

   (b) At what magnetic field strength was this spectrum obtained? How many spin states are available to a ^1H nucleus in such a field? How fast do ^1H nuclei precess when immersed in a field of this magnitude?

   (c) Was this spectrum obtained under cw or PFT conditions? How do you know?

   (d) What is the source of the signal at δ 0.00 ppm?

   (e) What is the chemical shift and relative intensity (expressed as the number of hydrogens represented) of the other signals in the spectrum?

   (f) Suggest a structure consistent with this spectrum, given that the molecular formula of the compound is $C_6H_{12}O_2$.

   (g) Assign all signals to hydrogens in your structure by calculating their predicted chemical shifts (use Table 6-1).

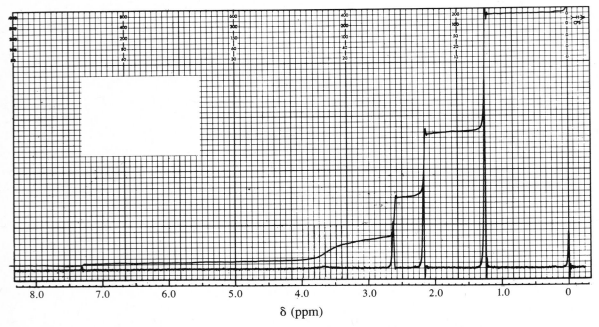

FIGURE I-1 The ^1H spectrum of $C_6H_{12}O_2$. © Sadtler Research Laboratories, Division of Bio-Rad Laboratories, Inc., 1974.

2. Consider the ^{13}C nmr spectrum shown in Figure I-2.

   (a) A ^{13}C nucleus possesses how many protons? Neutrons?

   (b) At what nominal frequency was this spectrum obtained?

   (c) What is the nuclear spin of a ^{13}C nucleus, and what is the importance of this number?

   (d) Was this spectrum obtained under cw or PFT conditions?

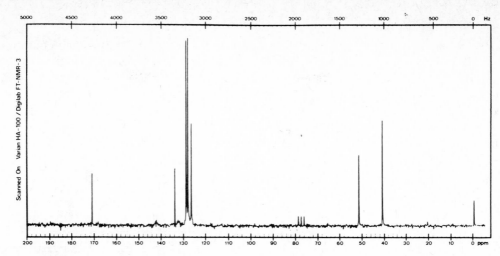

FIGURE I-2 The ^{13}C spectrum of $C_9H_{10}O_2$. © Sadtler Research Laboratories, Division of Bio-Rad Laboratories, Inc., 1983.

    (e) To what are the signals at δ 0.0, 76.0, 77.5, and 79.0 ppm due?

    (f) What is the chemical shift of each signal [other than those mentioned in part (e)]? What is the separation in Hz between the two most upfield signals?

    (g) Suggest a structure consistent with this spectrum, given that the molecular formula of the compound is $C_9H_{10}O_2$.

    (h) Assign all signals to carbons in your structure by calculating their predicted chemical shifts.

3. For each set of condensed spectral data below suggest a structure consistent with the data. Correlate observed chemical shifts with predicted values.

Molecular formula	Nucleus	Signal position (relative intensity)
(a) C_7H_8O	^1H	δ 2.43 (1H); 4.58 (2H); 7.28 (5H)[a]
(b) $C_3H_5Cl_3$	^1H	δ 2.20 (3H); 4.02 (2H)[a]
(c) $\begin{cases} C_{14}H_{12}O \\ C_{14}H_{12}O \end{cases}$	^1H ^1H	δ 3.88 (2H); 7.38 (10H)[a] δ 4.37 (2H); 7.20 (10H)[a]
(d) $C_{10}H_{14}O$	^{13}C	δ 31.5 (86); 33.7 (10); 114.9 (55); 125.9 (36); 141.6 (7); 154.7 (7)[b]
(e) C_5H_7N	^{13}C	δ 35.6 (18); 108.3 (88); 121.6 (58)[b]
(f) $C_7H_{12}O_2$	^{13}C	δ 25.5 (60); 25.9 (70); 29.0 (66); 43.1 (38); 183.0 (25)[b]

[a] *High Resolution NMR Spectra Catalog*, Varian Instruments, Palo Alto, 1963.
[b] *The Sadtler Guide to Carbon-13 NMR Spectra*, Sadtler Research Laboratories, Division of Bio-Rad Laboratories, Inc., Philadelphia, 1983.

4. Figures I-3a and b show the ^{1}H and ^{13}C nmr spectra of the same unknown compound.

(a) Without knowing the molecular formula, suggest a structure consistent with both spectra. [*Hint:* The molecular weight (the sum of the atomic masses of all atoms in the structure) of the compound is 116.]

(b) Assign each signal in both spectra to appropriate nuclei in your structure by calculating the predicted chemical shift of each.

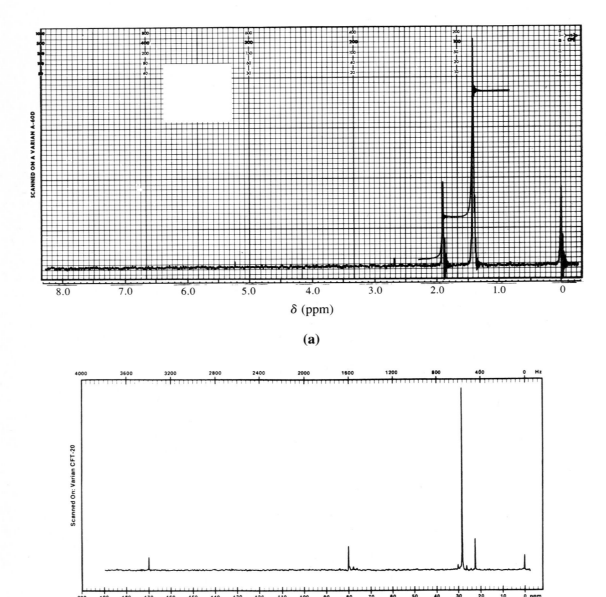

δ (ppm)

(a)

(b)

FIGURE I-3 (a) The ^{1}H spectrum of unknown compound with molecular weight 116 (from the *Sadtler Standard NMR Spectra*). **(b)** The ^{13}C spectrum of the same compound. © Sadtler Research Laboratories, Division of Bio-Rad Laboratories, Inc., 1974 and 1983, respectively.

8 FIRST-ORDER SPIN–SPIN COUPLING

THIS CHAPTER IS ABOUT

☑ **Unexpected Lines in an NMR Spectrum**
☑ **The ^{1}H Spectrum of Diethyl Ether**
☑ **Homonuclear ^{1}H Coupling: The Simplified Picture**
☑ **The Spin–Spin Coupling Checklist**
☑ **The $n + 1$ Rule**
☑ **Heteronuclear Spin–Spin Coupling**
☑ **Review Examples**

8-1. Unexpected Lines in an NMR Spectrum

Alas! Things are not as simple as they may have appeared. (Are they ever?) From the discussions in Chapters 6 and 7 you're probably under the impression that a typical ^{1}H nmr spectrum exhibits just one signal for each ^{1}H nucleus (or each set of equivalent ^{1}H nuclei), and that the same thing is true for ^{13}C spectra, as well as for spectra of any other isotope. Actually, this is not necessarily the case. Very often there are many more lines in a spectrum than we would have expected on the basis of symmetry alone. While these extra lines *do* make a spectrum more complex, they also offer valuable structural information that complements the chemical shift data. This chapter explains the source of these extra lines and shows how useful they can be for confirming the structures of molecules.

Take a moment to look back at the high-field (300-MHz) ^{1}H spectrum of toluene in Figure 5-4. Do you remember that there were more signals (at least six) for the aromatic hydrogens than the three we predicted from symmetry considerations? Similarly, in Figure 6-7 the two equivalent methine hydrogens gave rise to the six-line signal (area 2H) centered at δ 4.15 ppm (the sixth line is the barely discernable one at δ 4.40 ppm); the two equivalent methyl groups gave rise to a two-line signal (area 6H) centered at δ 1.25 ppm; and the two methylene hydrogens gave rise to a complicated-looking multi-line signal (area 2H) at δ 1.60 ppm. And do you recall from our discussion of the ^{13}C spectra in Chapters 5 and 7 that all these spectra involved use of a technique called proton spin decoupling? These phenomena are all related.

8-2. The ^{1}H Spectrum of Diethyl Ether

Before you look at Figure 8-1, try to predict the appearance of the ^{1}H spectrum of diethyl ether, CH_3—CH_2—O—CH_2—CH_3. By using symmetry and the data in Tables 6-1 and 6-2, you should have predicted two signals: one near δ 0.7 ppm for the six equivalent methyl hydrogens, and one near δ 3.0 ppm for the four equivalent methylene hydrogens. Now look at the 60-MHz ^{1}H spectrum in Figure 8-1. There *are* two signals, a three-line pattern centered at δ 1.13 ppm, and a four-line pattern centered at δ 3.38 ppm. And the ratio of the areas of these two complex signals is 6:4 (or 3:2). Let's examine these signals a little more closely.

The three-line methyl signal is called a **triplet**. The **multiplicity** of a signal equals the number of lines in the signal, so a triplet has a multiplicity of three. The three lines that constitute the triplet have an intensity ratio of approximately 1:2:1, and the two spacings between the lines are both 7 Hz (notice the Hz scale at the top of the spectrum). The chemical shift of the triplet is measured at its center, which corresponds to the middle line. The four-line pattern is called a **quartet** and has a multiplicity of four. Its

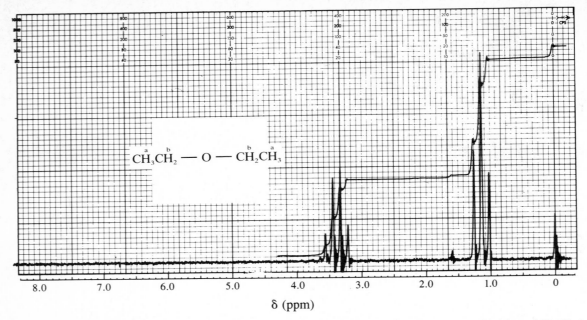

FIGURE 8-1 The 60-MHz ^{1}H spectrum of diethyl ether. ©Sadtler Research Laboratories, Division of Bio-Rad Laboratories, Inc., 1967.

chemical shift is also measured at its center, midway between the second and third lines. Again, the three spacings between lines are all 7 Hz, but the relative intensity of the lines is $1:3:3:1$. These intensity ratios, and the fact that the spacings in both the triplet and quartet are equal, are *not* accidental.

EXAMPLE 8-1 A **multiplet** is the generic term for any signal with two or more lines. How many lines constitute each of the following multiplets: doublet, quintet, sextet, septet?

Solution: Two, five, six, and seven, respectively.

note: A signal with only one line is called a **singlet**.

EXAMPLE 8-2 **(a)** What is the separation *in Hz* between the triplet and the quartet in Figure 8-1? **(b)** What would be the separation if the spectrum were run at 100 MHz?

Solution:

(a) You can either measure it directly from the Hz scale or calculate it from the difference in chemical shift. From Equation 5-1

$$\Delta v = (\Delta\delta)(v_0') = (3.38 - 1.13) \text{ ppm } (60 \text{ Hz ppm}^{-1}) = 135 \text{ Hz}$$

(b) Use a proportion of the type described in Sections 5-1B and 5-2:

$$\frac{\Delta v_1}{(v_0')_1} = \frac{\Delta v_2}{(v_0')_2}$$

$$\Delta v_2 = \Delta v_1 \left[\frac{(v_0')_2}{(v_0')_1}\right] = 135 \text{ Hz} \left(\frac{100 \text{ MHz}}{60 \text{ MHz}}\right) = 225 \text{ Hz}$$

In the last example, you found that the separation *between* the multiplets would increase from 135 Hz to 225 Hz if the spectrometer's operating (irradiation) frequency were increased from 60 to 100 MHz. That's no surprise. But here *is* the surprise: Even at 100 MHz, the separation between the lines of each multiplet is still 7 Hz! This spacing between the lines of a multiplet is called the (spin–spin) **coupling constant**, and is given the symbol J. Don't forget:

- While multiplet positions (and their differences) *in Hz* vary directly with frequency and field strength of the instrument, coupling constants (in Hz) are *independent* of these instrumental parameters.

When reporting nmr spectral data in **condensed form**, you should list the chemical shift (in ppm) of each multiplet, followed in parentheses by the type of multiplet (often abbreviated), coupling constant, and area (intensity).

EXAMPLE 8-3 Describe in condensed form the spectral data from the spectrum in Figure 8-1.

Solution: δ 1.13 (triplet, $J = 7$ Hz, 6H); 3.38 (quartet, $J = 7$ Hz, 4H).

8-3. Homonuclear ¹H Coupling: The Simplified Picture

I hope that by now your interest is piqued. Why do equivalent hydrogens give rise to multiplets in some cases and singlets in others? To understand this phenomenon, you must recall how the magnitude of the *net* (effective) magnetic field experienced by a nucleus establishes its precessional frequency, and thereby its chemical shift. If that's not second nature to you by now, perhaps you should review Section 6-1 before proceeding.

Let's examine the two hydrogen nuclei in one of the methylene groups of diethyl ether. Each of these hydrogens has two possible spin states (orientations, Section 2-2): aligned *with* the applied field ($m = \frac{1}{2}$) or *opposed* to the field ($m = -\frac{1}{2}$). These two spin states are approximately equally populated (Section 2-3), and nuclei in each state precess at the same frequency (Section 2-2). But the two hydrogens *together* can adopt four possible spin combinations: both *with* the field, both *opposite*, or one each, as shown in Figure 8-2. We can label each state by its M value, where M, the total magnetization, is the sum of the individual m values:

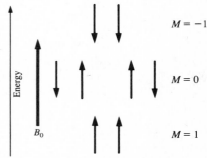

FIGURE 8-2 The spin states of two hydrogen (or other $I = \frac{1}{2}$) nuclei.

TOTAL MAGNETIZATION $M = m_1 + m_2 + m_3 + \cdots$ **(8-1)**

Notice that there is only one way to have both aligned *with* the field ($M = 1$), and one way for both to be *opposite* the field ($M = -1$), but *two* combinations where one is *with* and one is *opposite* ($M = 0$). Thus, two equivalent hydrogens together give rise to three spin states ($M = 1, 0, -1$) with population ratio of 1:2:1. Does this ratio ring a bell?

Now, for a moment, pretend you are one of the *methyl* hydrogens. (Well, at least I didn't ask you to be a naked proton!) Just three bonds away are located the two methylene hydrogens, which can be in any one of three possible spin states. Suppose they happen to be in the (slightly) more stable $M = 1$ state. Since their spins are aligned *with* the applied field, isn't it reasonable to expect that the net field you are experiencing is slightly *augmented* by their presence? If so, you will precess at a somewhat greater frequency, and your signal will be correspondingly shifted downfield. Alternatively, suppose the two neighboring hydrogens were in the $M = -1$ state. Their spins, being *opposite* to the applied field, would be expected to diminish your net field, causing you to precess more slowly and your signal to move upfield.

EXAMPLE 8-4 If the neighboring hydrogens were in the $M = 0$ state, what effect would they have on your precession rate?

Solution: None! When $M = 0$ the two methylene hydrogens are in opposite spin states, so their effects on the magnetic field cancel each other.

Okay, you can stop precessing now. But consider the implications of what we've just discussed. In a collection of diethyl ether molecules, the *methyl* hydrogens are next to *methylene* hydrogens that can be in any one of three spin states with relative probability 1:2:1. As a result, the *methyl* signal will be split into *three* lines with intensity ratio 1:2:1, as depicted in Figure 8-3. This phenomenon is aptly described by the term **spin–spin coupling** (or **splitting**). It is further characterized as **homonuclear coupling** because the coupling is between nuclei of the same isotope (here ¹H).

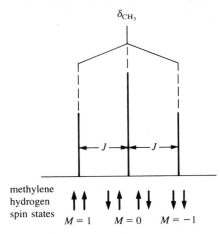

δ_{CH_3}

methylene hydrogen spin states

↑↑   ↓↑ ↑↓   ↓↓
$M = 1$ $M = 0$ $M = -1$

FIGURE 8-3 The methyl signal split into a triplet.

EXAMPLE 8-5 Now, consider what happens to the *methylene* signal as a consequence of the *three* neighboring (separated by three bonds) methyl hydrogens **(a)** How many spin states can *three* equivalent hydrogens adopt? Draw all possible combinations, and group them by M value. **(b)** What multiplicity do you predict for the methylene signal? What will be the intensity ratio of the lines? **(c)** Draw a diagram similar to Figure 8-3 for the methylene signal.

δ_{CH_2}

Solution:

(a) The three hydrogens together can adopt eight (2^3) possible spin combinations distributed among four spin states. Of these, one has all three against the field, three have two against and one with, three have one against and two with, and one has all three with the applied field.

↑↑↑   ↓↑↑ ↓↓↑   ↓↓↓
      ↑↓↑ ↓↑↓
      ↑↑↓ ↑↓↓

$M = \frac{3}{2}$ $M = \frac{1}{2}$ $M = -\frac{1}{2}$ $M = -\frac{3}{2}$

Methyl hydrogen spin states	M	Probability
↓↓↓	$-\frac{3}{2}$	1
↓↓↑ ↓↑↓ ↑↓↓	$-\frac{1}{2}$	3
↑↑↓ ↑↓↑ ↓↑↑	$\frac{1}{2}$	3
↑↑↑	$\frac{3}{2}$	1

FIGURE 8-4 The methylene signal split into a quartet.

(b) Quartet (four lines) in the ratio 1:3:3:1.
(c) See Figure 8-4.

Always remember that

• If nucleus A is coupled to nucleus B, the coupling constant (J) will have the *same magnitude* in the multiplets of *both* A and B.

8-4. The Spin–Spin Coupling Checklist

By now it should be increasingly apparent to you that the multiplicity (number of lines) of a given signal is related to the number of *neighboring* nuclei, *not* to the number of nuclei in the set giving rise to the signal. Let's examine this fact in a little more detail.

1. For two nuclei to engage in spin coupling, *both* must have nonzero nuclear spin ($I \neq 0$, Section 2-1). Certainly, if the neighboring nucleus had $I = 0$, it could adopt only *one* orientation in a magnetic field and therefore could not split the signal of the other nucleus.

2. To produce the effect of spin coupling, the coupling nuclei must be *nonequivalent*. We've already decided (from symmetry considerations) that all four methylene hydrogens are equivalent, as are the six methyl hydrogens. Why don't the three hydrogens within a given methyl group couple with each other? The answer is *they do*! But, for reasons we'll discover in the next chapter, we can never see the effects of this coupling because the methyl hydrogens are all equivalent.

3. Finally, you might wonder why the methyl hydrogens are coupled only to the two *nearest* (three bonds distant) methylene hydrogens, and not to the other two, located five bonds away. Here is the reason: The interaction between nuclear spins takes place predominantly through the intervening bonds (electrons). We'll discuss the factors that influence the magnitude of J in the next chapter, but for now we can offer the following generalization:

   • Normally, coupling is observed only when the nuclei are separated by no more than three bonds.

   To summarize, here is our spin-coupling checklist:

1. Are both nuclei magnetic ($I \neq 0$)?
2. Are they non-equivalent?
3. Are they separated by three or fewer bonds?

EXAMPLE 8-6 Which hydrogens in structure **8-1** will show coupling to each other?

$$
\begin{array}{c}
\quad\; \text{Br} \quad\; \text{I} \quad\;\; \text{Cl} \quad \text{H}_d \\
\quad\; | \quad\quad | \quad\quad | \quad\quad | \\
\text{Br—C—C—C—C—H}_d \\
\quad\; | \quad\quad | \quad\quad | \quad\quad | \\
\quad\; \text{H}_a \quad \text{H}_b \quad \text{H}_c \quad \text{H}_d
\end{array}
$$

8-1

Solution: By symmetry (or the lack of it), we see that H_a, H_b, and H_c are all nonequivalent, but the three methyl hydrogens (H_d) are equivalent. In the absence of coupling we would therefore expect *four* signals. Recalling that all ^1H nuclei have $I = \frac{1}{2}$, and counting the bonds separating them, we can complete our checklist:

Interaction	Number of intervening bonds	Coupling observable?
H_a–H_b	3	yes
H_a–H_c	4	no
H_a–H_d	5	no
H_b–H_c	3	yes
H_b–H_d	4	no
H_c–H_d	3	yes
H_d–H_d	2	no (equivalent!)

Thus, the only coupling we'll observe will be between H_a and H_b, H_b and H_c, and H_c and H_d.

8-5. The $n + 1$ Rule

Let's attack the general problem. How do we predict the number of lines and their relative intensity in a given multiplet, once we know the number of neighboring nuclei? We know that if there are *two* neighboring hydrogens, a *triplet* results (Figure 8-3), while if there are *three* neighboring hydrogens, a *quartet* results (Figure 8-4). A little thought should convince you that if there are n equivalent neighboring hydrogens, the signal will be split into $n + 1$ lines. This is because n hydrogens (or other nuclei with $I = \frac{1}{2}$) can generate only $n + 1$ different spin states from 2^n spin combinations.

EXAMPLE 8-7 (a) Diagram all possible spin combinations of four hydrogens. Arrange them into spin states by their value of M. (b) If another hydrogen were coupled to these four hydrogens, how many lines would constitute its multiplet and what would be their relative intensities?

Solution:

(a) There are sixteen (2^4) combinations distributed among five spin states:

M						Probability
-2			↓↓↓↓			1
-1		↓↓↓↑ ↓↓↑↓ ↓↑↓↓ ↑↓↓↓				4
0	↑↑↓↓ ↑↓↑↓ ↓↑↑↓ ↓↑↓↑ ↓↓↑↑ ↑↓↓↑					6
1		↑↑↑↓ ↑↑↓↑ ↑↓↑↑ ↓↑↑↑				4
2			↑↑↑↑			1

(b) Five lines (a quintet) with intensity ratio 1:4:6:4:1.

Now, review the intensity ratios we've encountered so far: triplet—1:2:1; quartet—1:3:3:1; quintet—1:4:6:4:1. Do these ratios sound at all familiar? Actually, they are the coefficients of the **binomial distribution**. There is a neat mnemonic device called **Pascal's triangle** that will enable you to generate these ratios in short order. A portion of it is shown in Table 8-1 (the triangle can be extended downward as far as you wish to go). Note how each number in the triangle is the sum of the two numbers diagonally above it. For example, each 10 in the last line is the sum of the 6 and 4 above it.

TABLE 8-1 Pascal's Triangle, the Coefficients of the Binomial Distribution

n	$n+1$			Intensity ratio				Multiplicity
0	1				1			singlet
1	2				1 1			doublet
2	3			1	2	1		triplet
3	4		1	3	3	1		quartet
4	5	1	4	6	4	1		quintet
5	6	1 5	10	10	5	1		sextet

EXAMPLE 8-8 What is the ratio of intenstities in a septet?

Solution: Generate the next line of Pascal's triangle to get 1:6:15:20:15:6:1.

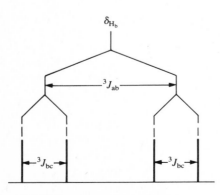

FIGURE 8-5 A doublet of doublets.

We should re-emphasize at this point that these ratios apply only to nuclei with $I = \frac{1}{2}$. Coupling to nuclei with $I > \frac{1}{2}$ is covered in the next section.

Let's return to our discussion of the ^1H spectrum of structure **8–1**. The signal for H_a will be split into a doublet by its one neighbor, H_b. Similarly, the methyl signal will be split into a doublet by H_c. But what about H_b? Its signal will be split by *both* H_a and H_c, but how? The answer depends on the *relative magnitude* of $^3J_{ab}$ versus $^3J_{bc}$ (the superscript 3 indicates the number of intervening bonds). Suppose for the moment that $^3J_{ab} > ^3J_{bc}$. The signal for H_b would be split into a doublet by H_a (with line separation $^3J_{ab}$), and each of the resulting lines would be further split into a doublet by H_c (with line separation $^3J_{bc}$).

The resulting pattern, called a **doublet of doublets**, is shown in Figure 8-5. Note that the intensity ratio of the lines is 1:1:1:1. If $^3J_{bc}$ were larger than $^3J_{ab}$, the pattern would look the same, except that the two coupling constants would be interchanged.

EXAMPLE 8-9 How would you describe the appearance of Figure 8-5 if the two coupling constants ($^3J_{ab}$ and $^3J_{bc}$) were accidentally equal?

Solution: The two middle lines would be superimposed, to give a *triplet* with line intensities 1:2:1 (see Figure 8-6).

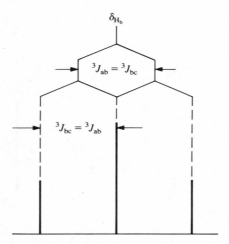

FIGURE 8-6 A doublet of doublets collapsed to a triplet.

As you can infer from Example 8-9, a triplet is really a doublet of doublets, where both coupling constants are equal. Similarly, a quartet can be seen as a doublet of doublets of doublets, with all three coupling constants equal.

But what about the signal for H_c? It will be split by both H_b and the three methyl hydrogens. Suppose that $^3J_{bc} > ^3J_{cd}$. Each line of the doublet due to coupling with H_b will be split further into a *quartet* by the three methyl hydrogens, as in Figure 8-7. This multiplet is called a *doublet of quartets*, and the line intensity ratio is 1:3:3:1:1:3:3:1.

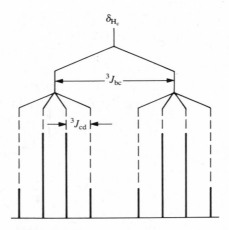

FIGURE 8-7 A doublet of quartets where $J_{bc} > J_{cd}$.

EXAMPLE 8-10 How would Figure 8-7 appear if (**a**) $^3J_{bc}$ were less than $^3J_{cd}$, or (**b**) if $^3J_{bc} = ^3J_{cd}$?

Solution:

(a) A *quartet of doublets* with intensity ratio 1:1:3:3:3:3:1:1, as shown in Figure 8-8a.

(b) A quintet with intensity ratio 1:4:6:4:1, as shown in Figure 8-8b.

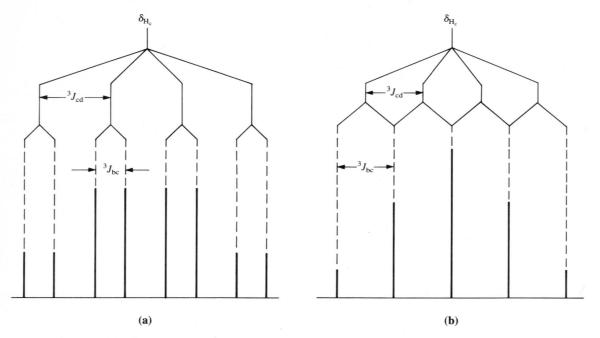

(a)   (b)

FIGURE 8-8 (a) A quartet of doublets where $J_{cd} > J_{bc}$; (b) a quintet where $J_{bc} = J_{cd}$.

We can generalize as follows:

1. If a nucleus is coupled to n other $I = \frac{1}{2}$ nuclei and all coupling constants are equal, the signal for the nucleus will be split into $n + 1$ lines whose intensity ratio can be predicted from Pascal's triangle.

2. If a nucleus is coupled to several sets of other $I = \frac{1}{2}$ nuclei (n_a of nucleus a, n_b of nucleus b, etc.), the total multiplicity (L) of the signal (assuming no overlap of any lines) is given by the product

TOTAL MULTIPLICITY ($I = \frac{1}{2}$) $L = (n_a + 1)(n_b + 1) \cdots = \prod_i (n_i + 1)$ (8-2)

where the $\prod_i$ indicates a product over i terms. Thus, in the case of H_c in structure **8-1** (with $n_b = 1$ and $n_d = 3$) there will be $(2)(4) = 8$ lines (barring accidental equivalence and overlap). The intensity ratio of the lines will depend on the relative magnitudes of the two coupling constants (compare Figures 8-7, 8-8a, and 8-8b).

EXAMPLE 8-11 (a) Predict in detail the appearance of the 100-MHz ^{1}H spectrum of the molecule with the following structure (**8-2**) and spectral parameters. Draw the spectrum on graph paper.

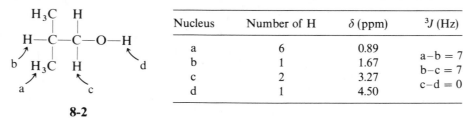

8-2

Nucleus	Number of H	δ (ppm)	^3J (Hz)
a	6	0.89	a–b = 7
b	1	1.67	b–c = 7
c	2	3.27	c–d = 0
d	1	4.50	

(b) Describe the resulting spectrum in condensed form.

Solution:

(a) Let's begin by transforming the δ values into Hz downfield from TMS by multiplying each by

100 Hz ppm^{-1} (Section 5-1). Thus, the H$_a$ signal is centered at 89 Hz, H$_b$ at 167 Hz, H$_c$ at 327 Hz, and H$_d$ at 450 Hz; mark these locations on your graph. The six equivalent methyl hydrogens (H$_a$) are coupled only to H$_b$ (why?). The H$_a$ signal therefore appears as a *doublet* with a separation between the lines of 7 Hz ($^3J_{ab}$). So, draw two equally intense lines (1:1 ratio, right?), one at 89 − 3.5 Hz, the other at 89 + 3.5 Hz. Remember that the total area of this doublet is six hydrogens, so each of these lines represents three hydrogens.

Let's shift our attention to H$_d$. It is not coupled to the two equivalent H$_c$'s ($^3J_{cd} = 0$), even though they're separated by only three bonds. This is because the OH hydrogens are undergoing rapid exchange (Section 6-7A), which erases the effects of coupling. (If, however, this exchange were slowed or stopped, coupling between H$_d$ and H$_c$ *would* cause the H$_d$ signal to be split into a triplet.) So draw a singlet at the appropriate position, one-third as intense as one of the methyl signals.

Since coupling between H$_d$ and H$_c$ can be neglected, the signal for H$_c$ is split into only a doublet ($^3J_{bc} = 7$ Hz) by H$_b$. Draw two lines, each as intense as the H$_d$ signal (since there are two H$_c$'s), one at 327 + 3.5 Hz, the other at 327 − 3.5 Hz.

Last but not least is H$_b$. Since it is coupled to two H$_c$'s and *six* H$_a$'s, will it appear as a triplet of septets (21 lines!) or a septet of triplets? Answer: Neither! Because both coupling constants ($^3J_{ab}$ and $^3J_{bc}$) are accidentally equal, H$_b$ will think itself coupled to *eight* equivalent nuclei. Therefore, its signal will be split into a *nine*-line (8 + 1) multiplet (nonet) with intensity ratio

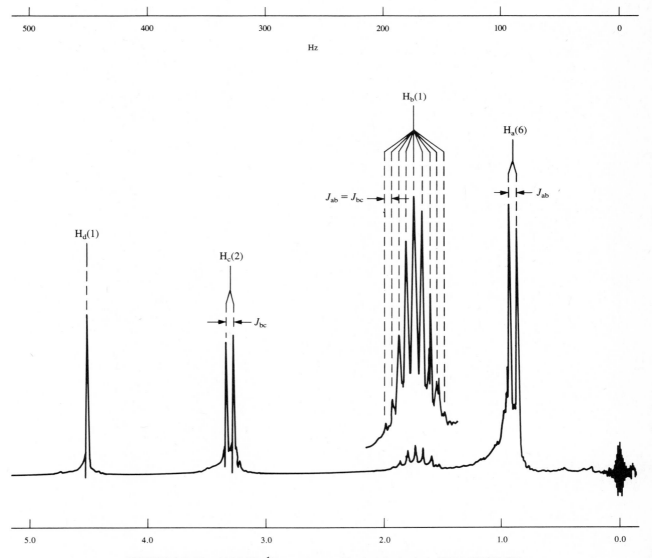

FIGURE 8-9 The 100-MHz ^{1}H spectrum of isobutyl alcohol, (CH$_3$)$_2$CHCH$_2$OH.

1:8:28:56:70:56:28:8:1. Furthermore, the *total* area of this multiplet will equal *one* hydrogen. Try to draw this on your diagram.

The actual spectrum of this compound is shown in Figure 8-9. How does it compare with your drawing? Notice how the nonet has to be amplified quite a bit to show all the lines. Incidentally, the reason that the TMS signal (δ 0.0) appears as a beat pattern is because TMS provided the *lock* signal as well as the reference signal for this spectrum.

(b) δ 0.89 (doublet, $J = 7$ Hz, 6H); 1.67 (nonet, $J = 7$ Hz, 1H); 3.27 (doublet, $J = 7$ Hz, 2H); 4.50 (singlet, 1H).

8-6. Heteronuclear Spin–Spin Coupling

Spin-coupling interactions are not limited to only hydrogens. Any magnetic ($I \neq 0$) nuclei within three bonds can do the same thing to a signal. Let's divide our discussion into nuclei where $I = \frac{1}{2}$, and all others.

A. Heteronuclear coupling between $I = \frac{1}{2}$ nuclei

From Chapter 2 we know that the common $I = \frac{1}{2}$ nuclei are ^{1}H, ^{13}C, ^{19}F, and ^{31}P. Consider the molecule H_2CF_2 (difluoromethane), and assume that the carbon is ^{12}C ($I = 0$):

$$\begin{array}{c} F \diagdown \quad \diagup H \\ C \\ F \diagup \quad \diagdown H \end{array}$$

The two hydrogens are equivalent, as are the two fluorines. Recalling that we can only examine one isotope at a time, can you predict the appearance of the ^{1}H spectrum of this compound? Because there are two neighboring (i.e., within three bonds) fluorines, the hydrogen signal is split into a *triplet* (intensity ratio 1:2:1), and the spacing between the lines will be $^2J_{HF}$. This kind of coupling is called **heteronuclear coupling** because it occurs between different isotopes or different elements.

EXAMPLE 8-12 Predict the appearance of the ^{19}F spectrum of difluoromethane.

Solution: The ^{19}F signal is also split into a *triplet* (intensity ratio 1:2:1) by the two equivalent hydrogens. And again, the line spacing is $^2J_{HF}$, the same value as in the ^{1}H spectrum.

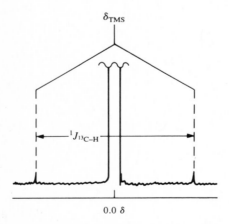

FIGURE 8-10 ^{13}C satellite peaks flanking a greatly amplified TMS signal.

But wait a minute! Why haven't we seen couplings between ^{1}H and ^{13}C in all our previous ^{1}H spectra? A good question! The answer is that ^{13}C constitutes only 1% of all carbon (Table 2-1). So only 1% of the hydrogens bonded to carbon are near a ^{13}C nucleus, and this small fraction of coupled hydrogens can normally be neglected. However, in some cases you *can* see the effect of ^{13}C–H coupling in ^{1}H spectra. Look at the highly amplified amplified singlet in the ^{1}H spectrum of TMS (Figure 8-10). Do you see the two small satellite peaks (sidebands) flanking the big signal? What are these? Your first answer might be spinning side bands (Section 3-2C). But suppose that changing the sample spinning rate doesn't affect their position. Then what would you say? [*Hint:* Each satellite peak represents $\frac{1}{2}$% of the main signal.]

If you guessed that the satellites are due to ^{13}C–H coupling, you're right! One percent of the methyl groups in a collection of TMS molecules possesses a ^{13}C atom. This atom splits the methyl hydrogen signal (the same one we use as our reference signal) into a *doublet*, with line spacing $^1J_{CH}$. Of course, you'll see these satellites only if you turn the signal amplification (gain) way up. But they *are* there, and the value of $^1J_{CH}$ is approximately 125 Hz, invariant with sample spin rate.

Ah, but this brings up an even bigger concern! Why aren't all the ^{13}C spectra we saw in Chapters 5 and 7 complicated by coupling between carbons and hydrogens? Alas, you caught me red-handed.

EXAMPLE 8-13 From your knowledge of heteronuclear coupling, predict the types of C–H couplings present in the *coupled* ^{13}C spectrum of

We can neglect $^{13}C-^{13}C$ coupling under natural abundance conditions because the probability of one molecule's having *two* ^{13}C nuclei is $(0.01)(0.01) = 0.0001$.

Solution: Let's prepare a coupling checklist.

Interaction	Separation (bonds)	Coupling observable?	Interaction	Separation (bonds)	Coupling observable?
C_a-H_a	1	yes	C_c-H_a	3	yes
C_a-H_b	2	yes	C_c-H_b	2	yes
C_a-H_c	3	yes	C_c-H_c	1	yes
C_a-H_d	4	no	C_c-H_d	2	yes
C_b-H_a	2	yes	C_d-H_a	4	no
C_b-H_b	1	yes	C_d-H_b	3	yes
C_b-H_c	2	yes	C_d-H_c	2	yes
C_b-H_d	3	yes	C_d-H_d	1	yes

Imagine how complex this spectrum would be! Why, the signal of C_a alone would be split into a quartet (by H_a) of doublets (by H_b) of triplets (by H_c)! Yet, look back at Figure 7-1. Why is it so simple, showing only a singlet for each carbon? Well, if you read the fine print in Chapters 5 and 7, you'll see the proviso that all ^{13}C spectra in those chapters were produced by means of a technique called **spin decoupling**. We'll discuss this technique again in Chapter 11, but for now we can say that spin decoupling allows us to "erase" the effect of coupling, greatly simplifying the appearance of the spectra.

EXAMPLE 8-14 Predict the multiplicity of each signal in the ^1H spectrum of the molecule represented by structure **8-3**.

8-3

Solution: First off, remember that ^{31}P *is* magnetic ($I = \frac{1}{2}$). The signal with the farthest downfield shift is that of H_d because of that hydrogen's similarity to a carboxylic acid hydrogen (near 12 ppm, Section 6-7A). Although H_d is only two bonds away from the phosphorus, coupling of H_d to P is not observed because of rapid exchange of the OH hydrogens (Section 6-7A). Vinyl hydrogen H_a's signal is split into a doublet (by phosphorus) of doublets (by H_b), but it is not coupled to H_d (why?). The most upfield signal, that of H_c, is split into a doublet (by phosphorus) of doublets (by H_b). The signal for H_b will be split into a doublet (by phosphorus) of doublets (by H_a) of triplets (by the two H_c's), for a total of twelve lines! We'll discuss this spectrum again in Section 9-10.

B. Heteronuclear coupling involving nuclei with $I > \frac{1}{2}$

In Section 5-4A (and Self-Test I, Problem 2e) we saw that $CDCl_3$, a very common nmr solvent, gives rise to *three* lines in its ^{13}C spectrum. Why is this so? Recall that D represents ^2H (deuterium), an isotope of hydrogen with $I = 1$, which can therefore adopt three $(2 \cdot 1 \cdot 1 + 1 = 3)$ orientations with essentially equal populations in a magnetic field. Thus, the *carbon* signal of $CDCl_3$ will be split into three equally intense lines.

Now, consider the molecule D_2CH_2. In this case each of the *two* ^2H nuclei can adopt three orientations $(m = 1, 0, -1)$ in a magnetic field (Section 2-2). Therefore, *two* ^2H nuclei can adopt... (can you figure it out?)... nine (3^2) spin combinations corresponding to *five* spin states $(M = 2, 1, 0, -1, -2)$. These are shown in Figure 8-11. Notice that the 1:2:3:2:1 probability ratio of these states (which is the same as the relative intensity of the multiplet lines coupled to them) is *different* from that predicted by Pascal's triangle, which only applies to $I = \frac{1}{2}$ nuclei. At any rate, the ^1H spectrum of D_2CH_2 shows a five-line pattern, with intensity ratio 1:2:3:2:1 and line spacing $^2J_{HD}$.

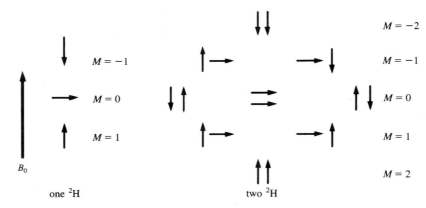

FIGURE 8-11 The possible spin states of one and two ^2H nuclei.

EXAMPLE 8-15 Predict the appearance of the deuterium (^2H) spectrum of D_2CH_2.

Solution: The (only) deuterium signal will be split into a triplet (1:2:1) by the two equivalent hydrogens. The line spacing $^2J_{HD}$ will be the same as in the ^1H spectrum.

EXAMPLE 8-16 Acetone-D_6 (structure **8-4**) is often used as a proton-free nmr solvent. It usually contains a small amount of acetone-D_5 (structure **8-5**). **(a)** Predict the appearance of the ^1H spectrum of acetone-D_5. **(b)** Predict the appearance of the ^{13}C spectrum of acetone-D_6.

8-4   **8-5**

Solution:

(a) There are two deuterium nuclei two bonds away and three others four bonds away from the sole hydrogen. Only the former ones will couple with the hydrogen. Therefore, the ^{1}H spectrum will exhibit a five-line pattern centered at δ 2.10 ppm, with intensity ratio 1:2:3:2:1.

(b) The carbonyl carbon signal will appear near δ 210 ppm (Table 7-5), and it will be split by the six equivalent deuterium nuclei (two bonds away) into $2 \cdot 6 \cdot 1 + 1 = 13$ lines. The two equivalent "methyl" carbons will appear near δ 30 ppm (Tables 7-1 and 7-2). Each will be split by a large one-bond coupling to the nearest three deuteriums, and by a smaller three-bond coupling to the three more remote deuteriums. Thus, the total multiplicity of this signal is $(2 \cdot 3 \cdot 1 + 1)(2 \cdot 3 \cdot 1 + 1) = 49$ lines! In actuality, for reasons we'll discuss in Chapter 9, only the one-bond coupling constant is large enough to be observed. Therefore, the carbonyl signal will be a singlet, and the methyl signal will have seven lines. (See Problem 4a in Self-Test II.)

We can now modify Eq. (8-2) into a more general form. The number of different spin states (different M values) available to n nuclei with nuclear spin I is $2nI + 1$, and the number of spin combinations is $(2nI + 1)^n$. Therefore, if no lines overlap, the total multiplicity, L, is given by

TOTAL MULTIPLICITY ($I \geq \frac{1}{2}$) $\qquad L = \prod_i (2n_i I_i + 1)$ $\qquad\qquad$ **(8-3)**

Notice how Eq. (8-3) reduces to Eq. (8-2) if we substitute a value of $\frac{1}{2}$ for I.

EXAMPLE 8-17 Predict the multiplicity of the ^{1}H signal of

$$\begin{array}{c} H \\ \diagdown \\ C \cdots F \\ \diagup \quad \diagdown F \\ D \end{array}$$

Assume that the carbon is ^{12}C ($I = 0$).
Solution: Let's construct a table:

Nucleus	n_i	I_i	$2n_i I_i + 1$
F	2	$\frac{1}{2}$	3
D	1	1	3

Therefore, $L = (3)(3) = 9$ lines! But remember: This "triplet of triplets" won't have the intensity ratios predicted by Pascal's triangle, because not all the nuclei have $I = \frac{1}{2}$.

8-7. Review Examples

Let's finish this chapter with two examples that review the concepts we've seen in the preceding sections.

EXAMPLE 8-18 (a) Suggest a structure for $C_{11}H_{12}O_2$, whose ^{1}H spectrum is shown in Figure 8-12, and whose ^{13}C spectrum (not shown) indicates the presence of an ester (Table 7-5). Be sure your structure accounts for all the observed multiplicities. (b) Measure from the spectrum the value of each coupling constant.

Solution:

(a) First, we notice a triplet, a quartet, a complex multiplet at δ 7.35 ppm, and two doublets (one centered at δ 6.33, the other at δ 7.73 ppm). The reason you can be sure that the line at δ 7.50 ppm is part of a doublet (as opposed to part of the complex multiplet) is because the spacing (J) between the two most deshielded lines exactly matches the spacing in the other doublet.

Next, we'll measure the integral (intensity) of each multiplet to determine the number of hydrogens each one represents.

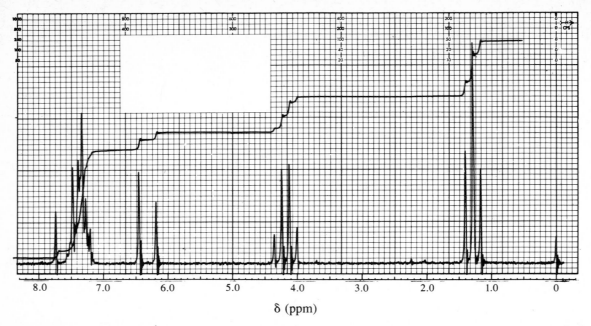

FIGURE 8-12 The 60-MHz ^{1}H spectrum of $C_{11}H_{12}O_2$. © Sadtler Research Laboratories, Division of Bio-Rad Laboratories, Inc., 1966.

δ (multiplicity)	Relative area	Number of H
1.30 (triplet)	10.3	3
4.18 (quartet)	6.8	2
6.33 (doublet)	3.4	1
7.35 (multiplet)⎤ 7.73 (doublet) ⎦	20.6	6
Totals 41.2		12

The overlapping multiplet/doublet must correspond to a 5H multiplet and a 1H doublet.

From the similarity of the triplet and quartet to those in Figure 8-1, we can confidently assign them to an —O—CH$_2$—CH$_3$ (*ethoxy*) group. The five-hydrogen multiplet is exactly where aromatic hydrogens appear, suggesting a monosubstituted aromatic ring. From their chemical shifts, the two doublets are most likely due to nonequivalent vinyl hydrogens coupled to each other. The only atoms not accounted for are one carbon and one oxygen, probably in the form of a carbonyl (C=O). We might propose two structures **A** and **B**:

A

B

The chemical shifts of the vinyl hydrogens (Table 6-3) are most consistent with structure **B**.

(b) From the Hz scale on Figure 8-12 we can measure most of the coupling constants. The triplet and quartet share a $^3J_{ab}$ value of 8 Hz, while the two doublets share a $^3J_{cd}$ value of 18 Hz. As we'll see in Example 9-8, the magnitude of $^3J_{cd}$ is very informative with regard to the exact structure of **B**. The aromatic hydrogen signal is complex because the couplings are not well-resolved. Although we will reconsider this type of aromatic system again in the next chapter, for now you can certainly appreciate that the five H_e's are actually divided among three sets (two ortho, two meta, and one para, Section 5-1A) that accidentally occur at nearly the same chemical shift. The meta hydrogens are coupled to both the ortho and para hydrogens, and there is even a small "long-range" interaction between the ortho and para hydrogens. You see how this would give a very complicated set of multiplets, except that they're compressed into a narrow region of the spectrum. It is impossible to measure these latter coupling constants directly from this spectrum, though there are alternative techniques for doing so. Be patient!

EXAMPLE 8-19 Examine the ^1H and ^{13}C spectra of the unknown compound $C_6H_4FNO_2$ shown in Figures 8-13a and b. (*note:* The ^{13}C spectrum is proton-decoupled.) (**a**) Suggest a structure for the compound. Account for the multiplicity of each signal in both spectra. (**b**) Measure from the two spectra the values of all coupling constants.

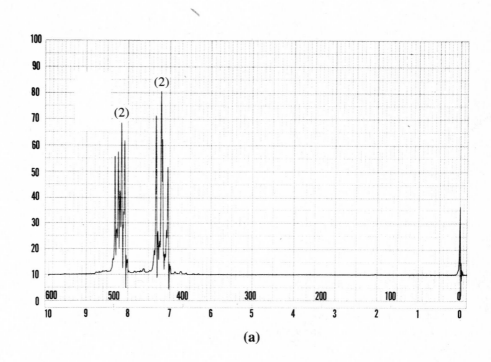

(a)

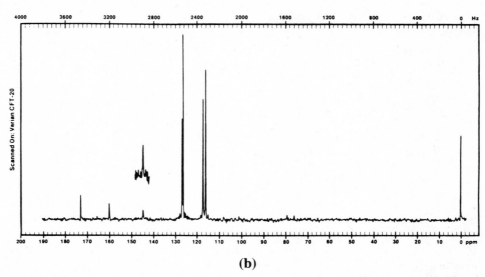

(b)

FIGURE 8-13 (**a**) The 60-MHz ^1H spectrum of $C_6H_4FNO_2$ (from *The Aldrich Library of NMR Spectra*); (**b**) the 20-MHz ^{13}C spectrum of the same compound. ©Sadtler Research Laboratories, Division of Bio-Rad Laboratories, Inc., 1983.

Solution:

(a) Beware! With fluorine present there are going to be extra coupling interactions! The ^1H spectrum exhibits two doublets of doublets (centered at δ 7.19 and 8.20 ppm), though the upfield one looks more like a triplet at first glance. Both are in the aromatic hydrogen region, and two of the four hydrogens can be assigned to each.

The ^{13}C spectrum shows seven lines for six carbons. The four lines from δ 115 to 128 ppm suggest an aromatic ring. Yet, if there are only six carbons in the structure, it is likely that all of them are part of the aromatic ring. This requires that the signals in the region δ 145–185 ppm also represent aromatic carbons, highly deshielded by substituent groups. A quick review of Table 7-4 shows that two of the most deshielding groups are F and NO_2. On the basis of the analysis so far we can suggest three isomeric structures:

ortho	meta	para

This table shows the predicted number of signals (before couplings are considered) for each structure:

Structure	^1H signals	^{13}C signals
ortho	4	6
meta	4	6
para	2	4

When you include both homonuclear and heteronuclear couplings, you can quickly see how the ortho and meta structures would give much more complicated spectra than those in Figure 8-13.

Let's focus our attention on the para isomer. The signal for H_a should occur (Table 6-4) at δ 7.29 ppm and should be split into a doublet (by F) of doublets (by H_b). The reason that H_a is split by only the H_b ortho to it (and not the H_b para to it) is because of the number of intervening bonds. Similarly, the H_b signal is predicted to appear at δ 8.24 ppm, split into a doublet by H_a. The fact that the actual signal appears as a doublet *of doublets* shows that there is also a long-range (four-bond) coupling between H_b and F.

Using the data in Table 7-4, we can compute the predicted chemical shifts of each carbon in the para isomer:

C	$\delta_{calculated}$
1	δ(benzene) $+ \Delta\delta$(F, α) $+ \Delta\delta$(NO$_2$, p) $= 128.5 + 34.8 + 5.8 = 169.1$
2	δ(benzene) $+ \Delta\delta$(F, o) $+ \Delta\delta$(NO$_2$, m) $= 128.5 - 12.9 + 0.9 = 116.5$
3	δ(benzene) $+ \Delta\delta$(F, m) $+ \Delta\delta$(NO$_2$, o) $= 128.5 + 1.4 - 4.8 = 125.1$
4	δ(benzene) $+ \Delta\delta$(F, p) $+ \Delta\delta$(NO$_2$, α) $= 128.5 + 20.0 - 4.5 = 144.0$

From these calculated values we must conclude that the two lines near δ 116 ppm are actually the two lines of a doublet for C_2. The same is true of the two lines centered at δ 126 ppm for C_3. The weak singlet at δ 145 ppm must correspond to C_4. This leaves the last two lines, which must constitute a doublet centered at δ 167 ppm for C_1.

(b) The $^{13}C-F$ coupling constants can be estimated by comparing the line spacings with the Hz scale at the top of the ^{13}C spectrum. The actual values determined by the computer when the spectrum was plotted are $^1J_{C_1-F} = 256$ Hz, $^2J_{C_2-F} = 24$ Hz, and $^3J_{C_3-F} = 11$ Hz. Note how the magnitude of the coupling constant decreases as the number of intervening bonds increases. More on this in Chapter 9.

In like manner, the H–H and H–F coupling constants can be estimated from the Hz scale on the ^1H spectrum. Remembering that $^3J_{ab}$ must equal $^3J_{ba}$, we look in the two multiplets for the spacings that match. That value, the spacing between the first and third (or second and fourth) lines of each multiplet is 9 Hz. The remaining spacings, between the first and second (or third and fourth) lines, are the H–F coupling constants. From the lower-field multiplet we extract a value of $^4J_{H_b-F} = 5$ Hz; from the upfield one, $^3J_{H_a-F} = 8$ Hz.

Now, if you're worried why a *four*-bond coupling was observed in this case, or why the lines of some multiplets deviate more than others from the intensities predicted from Pascal's triangle, you'll just have to go on to Chapter 9!

SUMMARY

1. If two nuclei in a molecule meet the criteria of the spin coupling checklist (below), the nmr signal for each nucleus will be split into multiple lines.

   Checklist
   (a) Are both nuclei magnetic ($I \neq 0$)?
   (b) Are they nonequivalent?
   (c) Are they separated by no more than three bonds?

2. The number of lines (L) in a given nmr signal is determined by the number (n) of **neighboring** coupled nuclei according to the the equation

$$L = \prod_i (2n_iI_i + 1) \tag{8-3}$$

3. In the case of coupling to n EQUIVALENT hydrogens (or other $I = \frac{1}{2}$ nuclei), the equation above reduces to $L = n + 1$.
4. Nmr signals consisting of multiple lines are characterized by the number of lines: doublet (2), triplet (3), quartet (4), etc. A signal with just one line is singlet. A signal with an unspecified number of lines is a multiplet.
5. The relative intensity of lines within a multiplet can be predicted from consideration of the various possible spin combinations of the neighboring nuclei. In the case of coupling to n equivalent hydrogens (or other $I = \frac{1}{2}$ nuclei), the relative intensities of the $n + 1$ lines are given by Pascal's triangle.
6. The separation between the lines of a multiplet is called the coupling constant J and is measured in hertz. The value of J is dependent only on the structural relationship between the nuclei, not on the field strength or frequency of the spectrometer. $^3J_{ab}$ indicates a three-bond coupling between a and b. Note also that $J_{ab} = J_{ba}$.

9 FACTORS THAT INFLUENCE THE SIGN AND MAGNITUDE OF J; SECOND-ORDER COUPLING EFFECTS

THIS CHAPTER IS ABOUT

☑ **Nuclear Spin Energy Diagrams and the Sign of *J***
☑ **Factors That Influence *J*: Preliminary Considerations**
☑ **One-Bond Coupling Constants**
☑ **Two-Bond (Geminal) Coupling Constants**
☑ **Three-Bond (Vicinal) Coupling Constants**
☑ **Long-Range Coupling Constants**
☑ **Magnetic Equivalence**
☑ **Spin System Designations**
☑ **Slanting Multiplets and Second-Order Coupling Effects**
☑ **Calculated Predictions of Spectral Appearance**
☑ **The AX → AB → A$_2$ Continuum**

9-1. Nuclear Spin Energy Diagrams and the Sign of *J*

Consider a situation where we have two nonequivalent hydrogens in a molecule, H$_a$ (with chemical shift δ_a) and H$_b$ (with chemical shift δ_b). For the sake of this discussion, we'll assume that $\delta_a > \delta_b$. Each of the ^{1}H nuclei can exist in either of two spin orientations (or states), and transitions between these spin states is what gives rise to nmr signals (Section 2-3). Figure 9-1 shows the relative energies of the four possible combinations of spin states of two ^{1}H nuclei. Recall that, since $\delta_a > \delta_b$, H$_a$ has the higher precessional frequency (ν_a) and the larger energy gap ($h\nu_a$) between its two spin states. In Figure 9-1 the four spin state combinations are numbered from the bottom up.

If asked how many transitions are possible among the four combinations of spin states, you might be tempted to list all of the following: $1 \to 2$, $1 \to 3$, $1 \to 4$, $2 \to 3$, $2 \to 4$, $3 \to 4$. But there is a so-called **selection rule** that controls the probability (and hence intensity) of each transition:

- An *allowed* transition involves only one nuclear spin flip; all transitions involving more than one flip are *forbidden*.

From the list of transitions we can therefore delete $1 \to 4$ and $2 \to 3$, because these involve the simultaneous flip of both spins. Transitions $1 \to 2$ and $3 \to 4$ result from the flipping of only the H$_b$ spin, and are thus responsible for the H$_b$ signal. Likewise, transitions $1 \to 3$ and $2 \to 4$ result from the flipping of only the H$_a$ spin, and give rise to the H$_a$ signal. Further, notice (from symmetry) that transitions $1 \to 2$ and $3 \to 4$ are *degenerate* (involve identical energy gaps), as are transitions $1 \to 3$ and $2 \to 4$.

Now, let's complicate the picture by including a coupling interaction between H$_a$ and H$_b$. This interaction will have one of two effects. Either the parallel spin states (1 and 4) will be raised in energy (by an amount Δ) and the antiparallel states (2 and 3) lowered by the same amount, or vice versa. This is easier to show in a diagram than to describe in words, so take a look at Figure 9-2. In Figure 9-2b are the same four (uncoupled) spin state combinations we saw in Figure 9-1. In this diagram, however, we've numbered the allowed transitions rather than the spin states themselves. So, transition $1 \to 2$ becomes transition 1, $1 \to 3$ becomes 2, $2 \to 4$ becomes 3, and $3 \to 4$ becomes 4. The spectrum that would result from these transitions is shown below the spin state diagram. Note how transitions 2 and 3 (which are degenerate) define δ_a, while transitions 1 and 4 define δ_b.

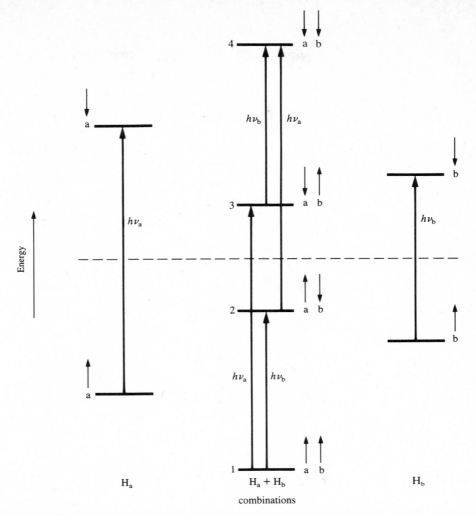

FIGURE 9-1 Possible combinations of spin states of two ^{1}H nuclei.

Figure 9-2a shows the spin state energies after the *parallel* states are raised by an amount Δ and the *antiparallel* ones are lowered by Δ. In Figure 9-2c the parallel states are *lowered* (by Δ) and the antiparallel ones are *raised* (by Δ). Careful inspection of the gaps between the energy levels in the two coupled cases reveals that transitions 1 and 4 are no longer degenerate, nor are transitions 2 and 3, as they are in the uncoupled case (b). In (a), transitions 3 and 4 have increased in energy gap (and hence frequency) by 2Δ, while transitions 1 and 2 have decreased by 2Δ. Therefore, the spectral lines resulting from transitions 1 and 2 move slightly upfield, while the lines corresponding to transitions 3 and 4 move an equal amount downfield. The resulting spectrum (shown below the spin state diagram) consists of two doublets, one centered at δ_a, the other at δ_b, with coupling constant J separating the two lines of each doublet.

EXAMPLE 9-1 What is the relationship between Δ and J?

Solution: Because the separation between spectral lines 2 and 3 in the coupled case is J Hz, each line must have moved $J/2$ Hz away from δ_a. Line 2 involves a transition between spin state 1 (which was *raised* by Δ) to spin state 3 (which was *lowered* by Δ). Therefore, $J/2$ must equal 2Δ, or $\Delta = J/4$.

Next, compare Figure 9-2c with 9-2b. Because this time the *antiparallel* states are raised (by Δ) and the *parallel* ones lowered, the positions of spectral lines 2 and 3 have been reversed, as have the positions of lines 1 and 4. But the resulting spectrum itself is indistinguishable from the one in Figure 9-2a.

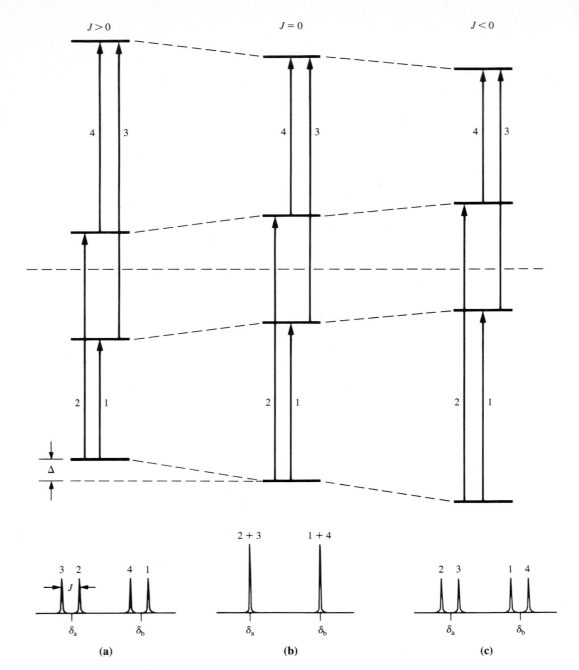

FIGURE 9-2 Effect of coupling on spin-state energies. **(a)** $J > 0$; **(b)** $J = 0$; **(c)** $J < 0$.

Well, I've kept you in suspense long enough. If the (a) and (c) (coupled) spectra are identical (except for the ordering of lines), what difference, if any, is there between the two cases? The answer is the *sign* of the coupling constant. A *positive* value ($J > 0$) implies that antiparallel spin states are lowered and parallel ones raised, while a *negative* value ($J < 0$) implies the converse. But here is the upshot of this whole discussion:

- While a coupling constant *does* possess a sign as well as a magnitude, whether the sign is positive or negative has *no* effect on the appearance of the spectrum. But the *relationship* between the signs of the coupling constants in a molecule (i.e., whether the signs are like or unlike) *can* have an effect when *second-order coupling* (Section 9-9) occurs.

Furthermore, when measuring J values from a spectrum, we can get only the *magnitude* of J, not its sign.

9-2. Factors That Influence *J*: Preliminary Considerations

A. Factors influencing the sign of *J*

From Chapter 8 we know that coupling constants have a magnitude that is somehow related to the number of intervening bonds. Furthermore, from the previous section we know that coupling constants have signs that indicate whether antiparallel spin states are lowered as a result of coupling ($J > 0$) or parallel states are lowered ($J < 0$). The *sign* of *J*, it turns out, is also sensitive to the number of intervening bonds.

These days theoretical chemists have quite a complete understanding of the factors that control the sign and magnitude of *J*. In fact, they can write out some pretty impressive equations that describe the spin–coupling interaction. But we'll try our best to steer clear of the high-powered math and speak in more qualitative terms.

Recall from Section 8-4 that the information about a nucleus' spin state is communicated to other nearby nuclei predominantly through the intervening bonding electrons. Thus, it comes as no surprise that, in general, the *magnitude* of *J* tends to decrease as the number of intervening bonds increases. A chemical bond consists of a pair of electrons occupying a **molecular orbital**, a region of space between the two nuclei. Electrons, like protons, have magnetic spin ($s = \pm\frac{1}{2}$), and in order for two electrons to occupy the same orbital, their spins must be *paired* (antiparallel, one with $s = \frac{1}{2}$, the other with $s = -\frac{1}{2}$, Section 2-1B). The simplest case of coupling between two hydrogen nuclei takes place in the hydrogen (H_2) molecule, as shown in Figure 9-3. Suppose that the preferred (more stable, lower energy) orientation of a nuclear spin is *opposite* to the spin of the nearest electron. Therefore, since the electron spins are paired, the two nuclear spins also prefer to be paired (antiparallel). And this is exactly the definition of a positive coupling constant:

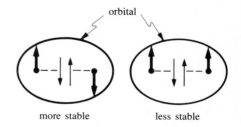

FIGURE 9-3 Nuclear spin orientations in an H_2 molecule. ↕ represents the magnetic moment of a nucleus, while ↑ represents the magnetic moment of an electron.

- The coupling constant between two nuclei is positive if the spins of the coupled nuclei prefer to be paired.

EXAMPLE 9-2 Using the spin-pairing model above, predict the sign of the two-bond coupling constant ($^2J_{HH}$) in H—^{13}C—H.

Solution: Building an extra bond onto the more stable sequence in Figure 9-3 gives a preferred parallel orientation of the two ^{1}H nuclei (Figure 9-4). Therefore, we predict that ^2J is negative.

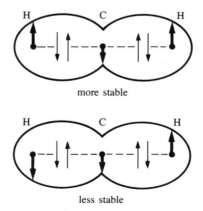

more stable

less stable

FIGURE 9-4 Nuclear and electronic spin orientations of a H—C—H group.

Although this model is rather naive, it does help us rationalize why one- and three-bond H–H, C–H, and C–C coupling constants are *usually* positive, while two-bond coupling constants are *usually* negative.

B. Factors influencing the magnitude of J

So much for the *sign* of J. But which factors determine its *magnitude*?

- All other things being equal, the magnitude of J is directly proportional to the product of the magnetogyric ratios (γ, Section 2-2A) of the two coupled nuclei.

EXAMPLE 9-3 (a) In the hydrogen molecule $^1J_{HH}$ is $+280$ Hz. What is the value (sign and magnitude) of $^1J_{HD}$ in H—D? (b) The 21.4-MHz ^{13}C spectrum of DCCl$_3$ (Self Test I, Problem 3e) consists of signals at δ 76.0, 77.5, and 79.0 ppm. Calculate $^1J_{CH}$ for HCCl$_3$.

Solution:

(a) The two values of γ (from Table 2-1) are 267.512×10^6 rad T^{-1}s^{-1} for ^{1}H and 41.0648×10^6 rad T^{-1}s^{-1} for D. Since the magnitude of J is directly proportional to the product of these values, we can set up a proportion:

$$\frac{^1J_{HD}}{^1J_{HH}} = \frac{\gamma_H\gamma_D}{\gamma_H\gamma_H} = \frac{\gamma_D}{\gamma_H}$$

Thus

$$^1J_{HD} = \frac{\gamma_D}{\gamma_H}(^1J_{HH}) = \frac{41.0648}{267.512}(280\ \text{Hz}) = 43\ \text{Hz}$$

And because this is a one-bond coupling constant, we can be confident that it is positive.

(b) These three lines are due to C–D coupling (Example 8-6b). To find the value of the corresponding C–H coupling constant, first find the value of the C–D coupling constant by applying Eq. (5-1):

$$^1J_{CD} = (\Delta\delta)(\nu_o)(79.0 - 77.5)\ \text{ppm}\ (21.4\ \text{Hz ppm}^{-1}) = 32.1\ \text{Hz}$$

Then use the same relationship as in part (a):

$$\frac{^1J_{CH}}{^1J_{CD}} = \frac{\gamma_{^{13}C}\gamma_H}{\gamma_{^{13}C}\gamma_D} = \frac{\gamma_H}{\gamma_D}$$

$$^1J_{CH} = \frac{\gamma_H}{\gamma_D}(^1J_{CD}) = \frac{267.512}{41.0648}(32.1\ \text{Hz}) = 209\ \text{Hz}$$

Several other factors are important in determining the magnitude of J, but we'll discuss these at the appropriate points in the next few sections. Also, you might want to take a peek at Problem 10, Self-Test II.

9-3. One-Bond Coupling Constants

The most commonly encountered one-bond coupling constant is $^1J_{CH}$. In fact, we saw an example of this in Section 8-6, when ^{13}C satellite peaks were mentioned. While the sign of all one-bond ^{13}C—H coupling constants is positive, the magnitudes depend on the nature of the orbital connecting the two nuclei.

The molecular orbital that constitutes a carbon–hydrogen bond results from the overlap of two **atomic orbitals**, one centered on the carbon and one on the hydrogen. (An *atomic* orbital is an orbital centered on a single nucleus.) All hydrogens use the same type of orbital (called a 1s orbital) to make their bonds. Carbon, on the other hand, uses a variety of different orbitals (s orbitals, p orbitals, and combinations of these called **hybrid orbitals**) to construct its bonds. The specific type of carbon orbitals used depends on the shape of the molecule. Although a detailed discussion of chemical bonding is beyond the scope of this book, Table 9-1 lists the types of hybrid carbon orbitals that occur in three common structures (single-, double-, and triple-bonded carbon), along with the corresponding value of $^1J_{CH}$. Note that an sp^n hybrid is one where the ratio of s-character to p-character (the relative contributions of s and p orbitals to the hybrid) is $1:n$, so the fraction of s-character (f_s) equals $1/(n + 1)$.

From the data in Table 9-1, we can see that $^1J_{CH}$ is directly proportional to the fraction of s-character

TABLE 9-1 The Effect of Carbon Hybridization on $^1J_{CH}{}^a$

Structure	Shape at carbon	C—C—H angle	Carbon hybrids	f_s	f_p	$^1J_{CH}$ (Hz)
	tetrahedral	109.5°	sp^3	0.25	0.75	125
	trigonal planar	120°	sp^2	0.33	0.67	156
H—C≡C—H	linear	180°	sp	0.50	0.50	249

^{a} Note that $f_s + f_p = 1$.

in the carbon hybrid:

$^1J_{CH}$ s-CHARACTER EFFECT $\qquad\qquad ^1J_{CH} = (500\ \text{Hz})(f_s)$ $\qquad\qquad$ **(9-1)**

The reason for this effect is that s orbitals interact more directly with the nucleus than do p orbitals. So, the greater a bond's s-character, the more nuclear spin information it communicates between neighboring nuclei.

EXAMPLE 9-4 What type of carbon hybrid is the C–H bond of H—CCl$_3$? [*Hint:* Look back at Example 9-3b.]

Solution: Solving Eq. (9-1) for f_s, we find

$$f_s = \frac{^1J_{CH}}{500\ \text{Hz}} = \frac{209\ \text{Hz}}{500\ \text{Hz}} = 0.418$$

Then, since $f_s + f_p = 1$,

$$f_p = 1 - f_s = 1 - 0.418 = 0.582$$

and

$$n = \frac{f_p}{f_s} = \frac{0.582}{0.418} = 1.39$$

Therefore, the hybrid orbital is an $sp^{1.39}$.

Many other homonuclear and heteronuclear one-bond coupling constants are known; a few of the more common ones are listed in Table 9-2.

TABLE 9-2 Some Representative One-Bond Coupling Constants^a

Type	^1J (Hz)
^{1}H—^{1}H	280
^{13}C—^{1}H	110–260
^{31}P—^{1}H	200–700
^{13}C—^{19}F	−280 to −350
^{13}C≡^{13}C	170
P—P	100–500
F—P=O	1000

^{a} As quoted in reference 1 at the end of this chapter.

EXAMPLE 9-5 From the value (Table 9-2) of $^1J_{C\equiv C}$ (170 Hz), devise a relationship that will enable you to predict the values of $^1J_{C=C}$ and $^1J_{C-C}$.

Solution: Because $^1J_{CC}$ is sensitive to the f_s value of *both* carbon orbitals, a relationship similar to Eq. (9-1) should prove valid for carbon–carbon coupling. A triple bond is formed in part by the overlap of two sp orbitals, a double bond results in part from overlap of two sp^2 orbitals, and a single bond involves two sp^3 orbitals. Thus, the desired relationship should include the *product* of both f_s values:

$^1J_{CC}$ **s-CHARACTER EFFECT** $\qquad\qquad ^1J_{CC} = (\text{constant})(f_s)^2 \qquad\qquad$ **(9-2)**

To evaluate the constant, put in the data for the triple bond:

$$^1J_{C\equiv C} = 170\ \text{Hz} = (\text{constant})(0.50)^2$$

$$\text{constant} = 170\ \text{Hz}/(0.50)^2 = 680\ \text{Hz}$$

Using this constant for $^1J_{C-C}$ and $^1J_{C=C}$ gives

$$^1J_{C-C} = (680\ \text{Hz})(0.25)^2 = 42.5\ \text{Hz}$$

and $\qquad\qquad\qquad ^1J_{C=C} = (680\ \text{Hz})(0.33)^2 = 75\ \text{Hz}$

The observed values are 35 Hz and 70 Hz, respectively.

9-4. Two-Bond (Geminal) Coupling Constants

In general, two-bond (geminal) coupling constants have a much smaller magnitude than one-bond coupling constants, and often (but not always) have negative signs. Geminal coupling constants are sensitive not only to magnetogyric ratios and s-character effects, but also to the angle between the bonds.

TABLE 9-3 Some Representative Two-Bond Coupling Constants[a]

Homonuclear Examples				Heteronuclear Examples	
Structure	^2J (Hz)	Structure	^2J (Hz)	Structure	^2J (Hz)[b]
$H_2C\big\langle^H_H$	−12.4	$(CH_2)_5\ C\big\langle^H_H$	−13	$^{13}C-C-H$	5
$H_2Si\big\langle^H_H$	+2.8	$(CH_2)_4\ C\big\langle^H_H$	−10.5	$H-C-F$	45
$H_2Sn\big\langle^H_H$	+15.3	$(CH_2)_3\ C\big\langle^H_H$	−9	$F-C-F$	160
$HFC\big\langle^H_H$	−9.6	$(CH_2)_2\ C\big\langle^H_H$	−4.3	$H-C-P\big\langle$	7–14
$HClC\big\langle^H_H$	−10.8	$CH_2{=}C\big\langle^H_H$	+2.5		
$HBrC\big\langle^H_H$	−10.2				
$HIC\big\langle^H_H$	−9.2				

[a] Data from references 1 and 2 at the end of this chapter.
[b] Absolute values.

Two-bond hydrogen–hydrogen coupling constants (as in H—C—H) usually fall in the range -9 to -15 Hz. Some typical values are listed in Table 9-3. In the cyclic molecules notice that as the ring gets smaller (and hence the H—C—H angle gets larger), the value of ^2J becomes more positive (less negative).

Perhaps you're wondering how two nuclei, if they are attached to the same atom, could possibly be nonequivalent, as required to produce observable coupling between them (Section 8-4). If you'll refer back to Section 4-3, you'll see one such case: diastereotopic nuclei! And even if the two *are* equivalent, there are tricks for measuring the coupling constant. For example, if two hydrogens are equivalent, one can (in principle, at least) substitute a deuterium for one of them, measure $^2J_{HD}$, then use Eq. (9-1) to calculate $^2J_{HH}$. Nonetheless, it *is* true that geminal nuclei often *are* equivalent, so that coupling between them does not complicate the spectrum.

9-5. Three-Bond (Vicinal) Coupling Constants

Of all the types of coupling, vicinal (three-bond) coupling can tell us the most about the three-dimensional arrangement of the atoms within a molecule. To understand why this is so, we'll have to define a new term: *dihedral angle.*

A **dihedral angle** (θ) is defined as the angle between two planes. If you open this book and lay it on your desk, the angle between the facing pages is 180°; if you close the book, the angle between the facing pages is 0° (Figure 9-5). In exactly the same way, vicinal bonds describe a dihedral angle, as also shown in Figure 9-5. The angle of interest is the one between the H_a—C—C plane and the C—C—H_b plane. In actual molecular structures, this dihedral angle can vary continuously (and rapidly) from 0 to 360° by rotation around the C—C single bond. Certain arrangements (conformations, Section 4-2), however, are more stable (and more prevalent) than others, and each one has its own specific dihedral angle. The fact that's most important in the present context is that the magnitude of ^3J (e.g., between H_a and H_b in Figure 9-5) varies with the dihedral angle. Martin Karplus was the first to describe this relationship by using an equation that now bears his name. The Karplus equation for vicinal H–H coupling, as modified by Bothner-By, has the form

KARPLUS RELATION
$$^3J_{HH} = (7 - \cos\theta + 5\cos 2\theta)\,\text{Hz} \qquad (9\text{-}3)$$

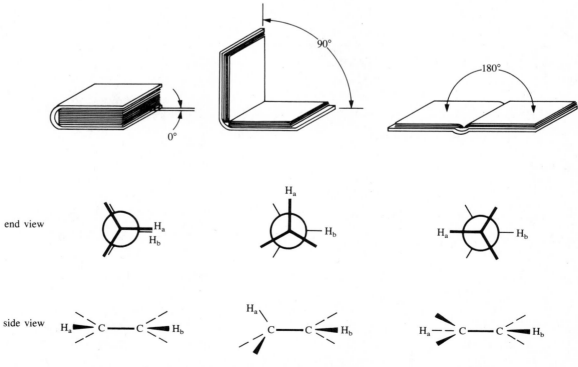

FIGURE 9-5 Dihedral angles.

Figure 9-6 is a graph of this equation. Notice how J reaches its maximum values at $0°$ ($J = 11$ Hz) and $180°$ ($J = 13$ Hz) and its minimum value at $90°$ ($J = 2$ Hz). This is because the interaction between the two vicinal orbitals (bonds) is at its maximum at $0°$ or $180°$ and decreases to nearly zero at $90°$.

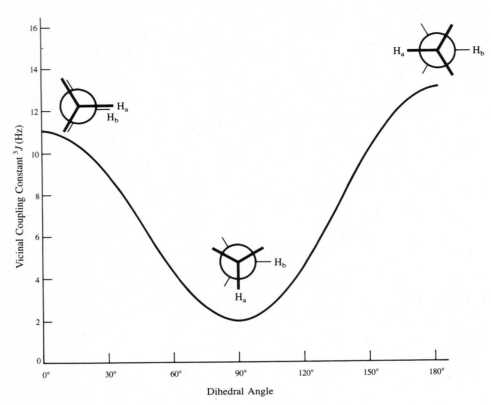

FIGURE 9-6 The Karplus relationship for vicinal coupling.

EXAMPLE 9-6 Cyclohexane and its derivatives usually exist in nonplanar conformations with the attached hydrogens occupying sites that are either *axial* (a) or *equatorial* (e).

Predict the values of ^3J for H_x–H_e and H_x–H_a and of ^2J for H_a–H_e.

Solution: From the structural diagrams it is clear that the H_x–H_e dihedral angle is $60°$, while the H_x–H_a angle is $180°$. Using Eq. (9-3) (or Figure 9-6),

$$^3J_{xe} = 7 - \cos 60° + 5\cos 120° = 4 \text{ Hz}$$

$$^3J_{xa} = 7 - \cos 180° + 5\cos 360° = 13 \text{ Hz}$$

To get the value of ^2J (a *geminal* coupling constant), refer back to Table 9-3, which gives the value of $^2J_{HH}$ for cyclohexane as -13 Hz. By the way, cyclohexane is another example in which two hydrogens attached to the same carbon (H_e and H_a) are *non*equivalent.

As mentioned above, in most molecules possessing single bonds there is a constant and rapid (on the nmr time scale) interconversion of all possible conformations, so the dihedral angles vary among all possible values. Since each conformation has a different value of the vicinal coupling constants, the observed values of ^3J represent a weighted average over all conformations. Such coupling constants can also show a dependence on temperature, because the ratio of the populations of the various conformations is a function of temperature.

EXAMPLE 9-7 Suppose a molecule were limited to just two conformations, A (where $^3J = 13$ Hz) and B (where $^3J = 4$ Hz). Furthermore, suppose there is rapid interconversion between these two conformations and that the observed value of ^3J is 7 Hz. Calculate the average fraction of the molecular population in each conformation.

Solution: Let f_A equal the fraction of A. Therefore, $f_B = 1 - f_A$. Since the observed coupling constant is the weighted average over both conformations, we can write

$$^3J_{obs} = 7 \text{ Hz} = f_A(13 \text{ Hz}) + f_B(4 \text{ Hz})$$

$$= f_A(13 \text{ Hz}) + (1 - f_A)(4 \text{ Hz})$$

Thus, $f_A = 0.33$ and $f_B = 0.67$.

Vicinal coupling across double bonds shows a similar variation with structure. But the presence of the double bond has two important consequences. First, the double bond does not allow rotation, and therefore limits vicinal relationships to either 0° (cis) or 180° (trans).

cis   trans

Second, because of the extra intervening bond, there are more electrons communicating spin information between the nuclei, and ^3J is about 20% greater than in the case of three single bonds.

EXAMPLE 9-8 Refer back to Example 8-18. In reality, there are two closely related structures for B: one cis, one trans. From the value of $^3J_{cd}$ (18 Hz), which structure best fits the spectral data?

B (trans)   **B′** (cis)

Solution: The H—C=C—H dihedral angle is 180° in **B**, 0° in **B′**. The 18-Hz coupling constant is evidence for structure **B**. Structure **B′** would exhibit a smaller value (about 10 Hz).

It is somewhat anomalous that the value of $^3J_{HH}$ in the linear molecule acetylene (H—C≡C—H) is only 9 Hz, because both the presence of the triple bond and s-character effects might have led us to predict a larger value.

So far in this section we've dealt only with homonuclear (^{1}H) vicinal coupling. But it should come as no surprise that the same type of Karplus angular dependence is seen for heteronuclear vicinal coupling constants as well. Table 9-4 lists some representative values.

TABLE 9-4 Typical Heteronuclear Vicinal Coupling Constants[a]

Type	^3J (Hz)	Type	^3J (Hz)
$^{13}C-C-C-H$	5	F,C=C,F	−120
$F-C-C-H$	5–20	$P-C-C-H$	13
H,C=C,F	40	P,C=C,H	30
C=C (H, F)	18	P,C=C,H	50
$F-C-C-F$	−3 to −20	$P-O-C-H$	5–15
F,C=C,F	30–40		

[a] Taken mainly from reference 1 at the end of this chapter.

9-6. Long-Range Coupling Constants

As mentioned in Section 8-4, coupling interactions are normally observed only when the nuclei are separated by no more than three bonds. (A multiple bond counts as only one bond separation.) Yet, there are special cases (e.g., Example 8-19) where coupling is observed over four, five, or even more bonds! Such **long-range coupling** can be anticipated under the following two circumstances:

1. There are one or more multiple bonds between the coupling nuclei, or
2. The molecule is rigid and the two nuclei possess a so-called **W spatial relationship**.

Electron pairs involved in multiple (i.e., double or triple) bonds behave somewhat differently from those in single bonds. Not only are there differences in geometry, hybridization, and *s*-character (Section 9-3), but there is also considerable evidence that multiple bond electrons are easier to *delocalize* (spread out) through the other molecular orbitals in a molecule. This delocalization of the electrons in multiple bonds was previously encountered as resonance or conjugation in Section 7-4. And because of this delocalization, multiple-bond electrons can communicate nuclear spin information over more than just three bonds. Several examples of this type of long-range coupling are listed in Table 9-5.

The other situation where long-range coupling can be expected is when the orbitals connecting the two coupling nuclei are forced by a rigid molecular structure to adopt a W relationship:

In such a structure it is believed that nuclear spin information is communicated between nuclei X and Y by overlap of the "tails" of the C—X and C—Y orbitals. This overlap "short-circuits" or sidesteps the other two intervening C—C bonds, making this type of long-range interaction resemble a three-bond coupling. But again, the molecular structure must constrain these bonds into this W relationship for long-range coupling of this sort to be important. For example, compare the value of the four-bond $H_e-H_{e'}$ (W relationship) coupling constant in cyclohexane (Table 9-6) with the negligibly small values of the four-bond $H_e-H_{a'}$ and $H_a-H_{a'}$ (not W) coupling constants.

The bottom line of this section is that long-range coupling (over more than three bonds) *is* possible, but only under these two circumstances. And even then, it usually involves relatively small coupling constants.

TABLE 9-5 Long-Range Coupling Mediated by Multiple Bonds[a]

Interaction	Number of bonds (n)	^nJ (Hz)
H—H_o	3	8
H—H_m	4	2
H—H_p	5	0.5
F—H_o	3	8
F—H_m	4	7
F—H_p	5	2
H—C—C=C—C—H	5	1–2
H—C=C—C—H	4	−2
H—C—C=C—C=C—H	5	±1
H—C—$C\equiv C$—H	4	−2
H—C—$C\equiv C$—$C\equiv C$—C—H	7	1
H—C—C—C—F	4	5
H—C=C=C—H (allene)	4	6
H—C=C=C—C—H	5	3

[a] Taken mainly from reference 1 and 3 at the end of this chapter. (See Problem 7 in Self-Test II.)

TABLE 9-6 Long-Range Coupling Mediated by W-Type Overlap

	Interaction	^4J (Hz)
	ee'	+1.8
	ea' (ae')	−0.4 $\left.\right\}$ *not* "W"
	aa'	−0.9
	H—H	1
	H—H	7

9-7. Magnetic Equivalence

In Chapter 4 we encountered the concept of equivalence: If two (or more) nuclei are related by virtue of an axis or plane of symmetry, they are said to be symmetry- (or chemically) equivalent. Furthermore, chemically equivalent nuclei precess at exactly the same frequency, and hence give rise to one nmr signal (coupling notwithstanding).

There is another kind of equivalence we need to introduce at this point. To do so, let's examine the structure discussed in Example 8-19, *p*-fluoronitrobenzene (structure **9-1**).

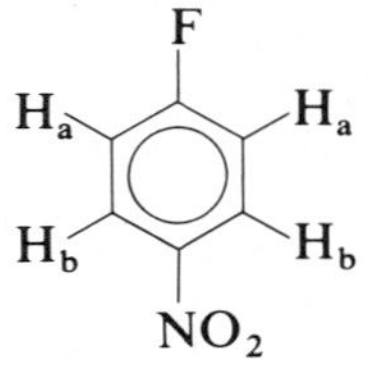

9-1

The two H_a's are chemically equivalent, being related by both a plane of symmetry and a C_2 axis; so are the two H_b's. Yet, as we found in Example 8-19, a given H_a couples only with the H_b ortho to it (three bonds away), and *not* to the other H_b (five bonds away). Thus, as far as an H_a is concerned, the two H_b's are *not* equivalent, even though *we* know they *are* equivalent. How can we resolve this paradox?

We describe this situation by saying that the two H_b's are *chemically equivalent* (i.e., they are symmetry-equivalent and occur at the same chemical shift), but they are *not* **magnetically equivalent**. For two (or more) nuclei to be magnetically equivalent, they must not only be chemically equivalent but also be equally coupled to any other nucleus. Thus, all *magnetically* equivalent nuclei are also *chemically* equivalent, but not all *chemically* equivalent nuclei are *magnetically* equivalent. We'll use a prime to indicate when two nuclei are chemically but *not* magnetically equivalent. So, we can relabel structure **9-1** as follows:

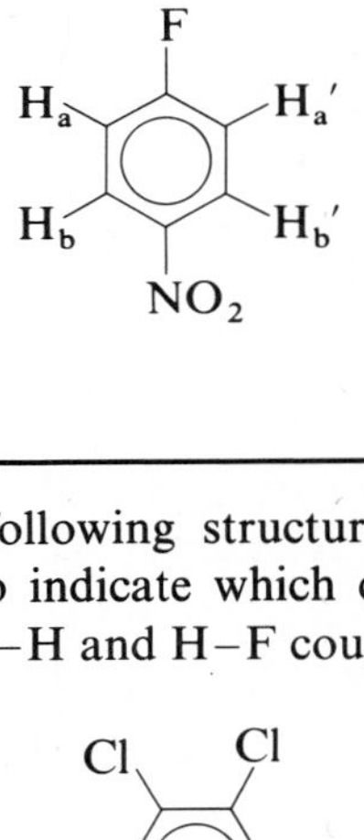

EXAMPLE 9-9 The subscripts in the following structures indicate which nuclei are chemically equivalent (by symmetry). Use primes to indicate which chemically equivalent nuclei are *not* also magnetically equivalent. Consider only H–H and H–F coupling.

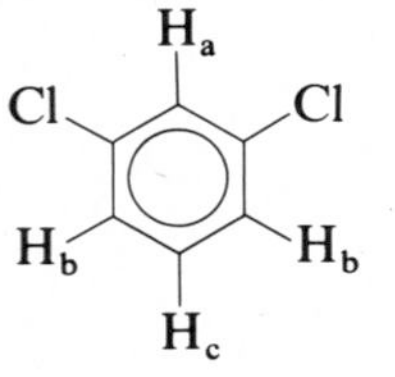

A 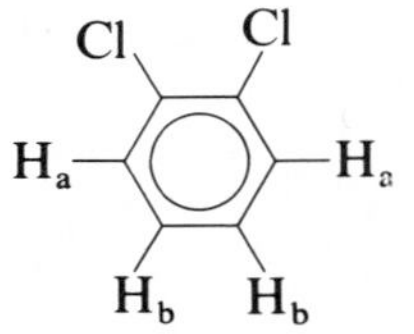   **B** 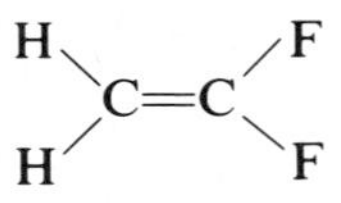   **C**

Solution:

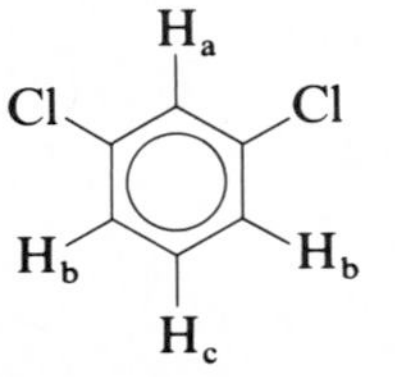

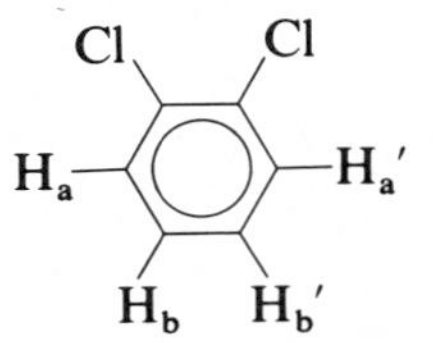

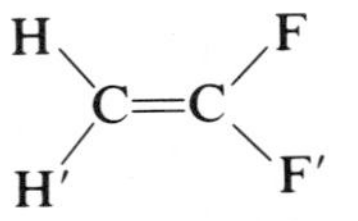

In **A**, the H_b's *are* magnetically equivalent (no prime) because their couplings to H_c are equal and their couplings to H_a are equal. In **B**, the H_a's are *not* magnetically equivalent, nor are the H_b's, because ortho (three-bond) coupling between H_a and H_b is different from (larger than) the meta (four-bond) coupling. And in **C**, the hydrogens are *not* magnetically equivalent, nor are the fluorines, because $^3J_{trans}$ is greater than $^3J_{cis}$.

The situation becomes slightly more complicated when we consider molecules with multiple conformations. For example, look carefully at the structure of 1,1,1-trifluoroethane (**9-2**). If this molecule were "frozen" in the conformation shown, the hydrogens would *not* be magnetically equivalent because

side view   **9-2**   end view

their couplings to the three fluorines involve three different dihedral angles and hence three different values of $^3J_{HF}$. Thus, we'd be tempted to label them as H, H', H'', F, F', F''. However, just as we saw with the methyl group of toluene (Section 4-2), there is rapid rotation around the C—C single bond, which averages all the H–F couplings (see Example 9-7). Operationally, therefore, all three hydrogens are rendered magnetically equivalent by the rapid rotation, as are the three fluorines. But if this rotation were slowed or stopped, the effects of their magnetic *non*equivalence would re-emerge.

9-8. Spin System Designations

We're now ready to introduce a shorthand method of labeling spin-coupled systems that will facilitate recognition of the associated spectra. Assign a capital letter to each magnetic nucleus involved in the coupled system. If two (or more) nuclei are chemically equivalent, give them the same letter. If two (or more) chemically equivalent nuclei are *not* also magnetically equivalent, distinguish them by labeling one with prime (or multiple primes). Finally, make the assigned letters reflect the differences among the chemical shifts of the various nuclei. Thus, for two nuclei of very different chemical shift (or different isotopes), pick A and X; for nuclei that have very similar (but not identical!) chemical shifts, use A and B. Just how much of a difference constitutes "very different" or "very similar" is discussed in the next section.

Refer back to the structure of 1,1,1-trifluoroethane (**9-2**). Assuming rapid bond rotation, we would designate it as an A_3X_3 spin system, where A_3 represents the three magnetically equivalent hydrogens and X_3 the three magnetically equivalent fluorines. Of course, if bond rotation were stopped, the system would become an AA'A''XX'X'' system, because the magnetic equivalence would have been lost.

EXAMPLE 9-11 Label the H/F spin system (neglecting carbon coupling) in *p*-fluoronitrobenzene, coupling to carbon.

Solution: **A**: AB_2C (all chemical shifts are similar); **B**: AA'BB'; **C**: AA'XX'.

EXAMPLE 9-11 Label the H/F spin system (neglecting carbon coupling) in *p*-fluoronitrobenzene, structure **9-1**.

Solution: The completely correct answer is AA'BB''X. But since para (five-bond) couplings (A–B' and A'–B) are essentially zero, we can regard the spin system as comprising two equivalent (and superimposed) ABX systems.

But why go to all the trouble of labeling spin systems? The answer is that

- Similar spin systems give similar spectra, and we can learn to recognize these recurring spectra and know immediately what type of structure generates them.

For example, an ethoxy group (CH_3—CH_2—O—) is an example of an A_3X_2 (or A_3M_2) spin system and always gives the familiar triplet (for A_3 split by X_2) and quartet (for X_2 split by A_3), as we saw in Section 8-1 and Example 8-18.

EXAMPLE 9-12 What type of spectra would you expect from the following spin systems: (a) AX; (b) AA'XX'; (c) AM_2X?

Solution:

(a) Two doublets, one for A (split by X), and one for X (split by A).

(b) The A signal will be split into a doublet of doublets by X and X'; the X signal will also be split into a doublet of doublets by A and A'.

(c) The A signal will be split into a triplet (by the two M's) of doublets (by X); the M signal will be split into a doublet (by A) of doublets (by X), while the X signal will be split into a triplet (by the two M's) of doublets (by A).

9-9. Slanting Multiplets and Second-Order Coupling Effects

If you were very observant, you might have noticed a recurring feature in the coupled ^1H spectra we discussed in Chapter 8. Take a moment to re-examine the spectra in Figures 8-1, 8-9, 8-12, and 8-13a and pay particular attention to the intensity ratios within each multiplet. Can you see the trend?

For example, in Figure 8-1 we would expect the quartet to exhibit (from Pascal's triangle) an intensity ratio of 1:3:3:1, while the triplet should have a ratio of 1:2:1. Although the observed ratios are indeed very close to these expectations, there *is* an unmistakable trend: the higher-field lines of the quartet are slightly more intense than the lower-field lines, while the lowest-field line of the triplet is slightly more intense than the highest-field line. Another way of saying this is that the *inner* lines of both multiplets (those lines closer to the other coupled multiplet) are more intense than the *outer* lines. This asymmetry of the multiplets, sometimes referred to as slanting, canting, or leaning of the multiplets, is the result of so-called **second-order effects**. Such effects can either help or hinder the interpretation of complex spectra.

Here is a way that multiplet slanting can be helpful. Look back at Figure 8-9. Notice how the doublets (for H_a and H_c) are slanted toward the multiplet of H_b, with which both H_a and H_c are coupled. The same is true in Figure 8-13A, where the two doublets of doublets slant toward each other. Thus, slanting helps us know where to look for the other coupled multiplet(s).

But look at the two doublets in Figure 8-12. Here, the departure from the expected intensity ratios of 1:1 is so great that an uninitiated person might not recognize them as doublets at all. (Aren't you glad you're not in that category?)

Why is the asymmetry very slight in some cases and much more pronounced in others? To begin to answer that question, we need to define a new (well, not really new) quantity, $\Delta v/J$, where Δv is the difference in chemical shift (measured in Hz) between the two multiplets, and J (in Hz) is the absolute value (magnitude without sign) of the coupling constant they share.

EXAMPLE 9-13 Referring again to Figure 8-12, calculate $\Delta v/J$ for (a) the pair of doublets, and (b) the quartet/triplet.

Solution:

(a) From the solution to Example 8-18, we know that the two doublets are centered at δ 6.33 and 7.73 ppm and that they share a ^3J value of 18 Hz. The spectrum was run at 60 MHz; therefore

$$\Delta v = (7.73 - 6.33) \text{ ppm } (60 \text{ Hz ppm}^{-1})$$

$$= 84 \text{ Hz}$$

Thus

$$\frac{\Delta v}{J} = \frac{84 \text{ Hz}}{18 \text{ Hz}} = 4.7$$

(b) Again using the data in the solution to Example 8-18,

$$\frac{\Delta v}{J} = \frac{(4.18 - 1.30) \text{ ppm} (60 \text{ Hz ppm}^{-1})}{8 \text{ Hz}} = 21.6$$

From Example 9-13, you can infer that the extent to which the intensity ratios depart from first-order (Pascal's triangle) expectations is a function of $\Delta v/J$. If this ratio is large (say, > 10), we describe the spin system as a **weakly coupled system**, and the resulting multiplets will exhibit essentially first-order intensity ratios (as do the triplet and quartet in Figure 8-12). But as $\Delta v/J$ decreases, second-order effects (such as multiplet slanting and even the appearance of "extra" lines) become increasingly important. Such a system is said to be a **strongly coupled system**.

In Chapter 5 we discussed the effect of increasing the operating frequency (and field strength) of our nmr spectrometer. You can now appreciate another advantage of high-field instruments. You know from Sections 2-2 and 5-2 that Δv (in Hz) increases *linearly* with field strength, while (from Section 8-2) J remains unaffected. Thus, the *ratio* $\Delta v/J$ also increases linearly with field strength, and therefore

- Coupled spectra become more first-order in appearance as the field strength increases.

As an example of this effect, compare the 60-MHz (1.41-T) ^{1}H spectrum of I—CH$_2$—CH$_2$—CO$_2$H (Figure 9-7) with its (partial) 300-MHz (7.04-T) spectrum (Figure 9-8). Notice how the signals for the coupled CH$_2$ groups are transformed from a strange (but symmetrical) multiplet at 60 MHz (where $\Delta v/J = 2.1$) into a pair of well-resolved triplets (as expected from first-order considerations) at 300 MHz ($\Delta v/J = 10.8$). From Figure 9-7 you can also see that in strongly coupled systems (that is, where $\Delta v/J$ is small), not only are the multiplets slanted, but extra lines also appear in the multiplets. We'll talk more about this in the next two sections, but for now remember:

- Strongly coupled systems lead to more complicated spectra than first-order rules (Chapter 8) predict.

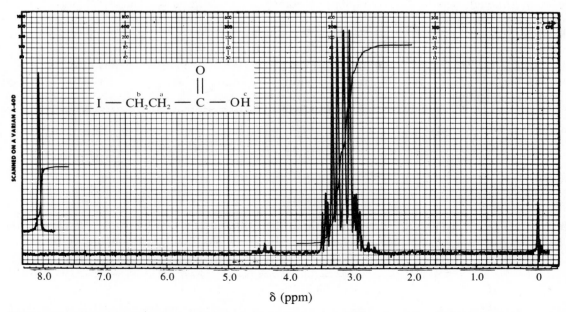

FIGURE 9-7 The 60-MHz ^{1}H spectrum of I—CH$_2$—CH$_2$—CO$_2$H. © Sadtler Research Laboratories, Division of Bio-Rad Laboratories, Inc., 1973.

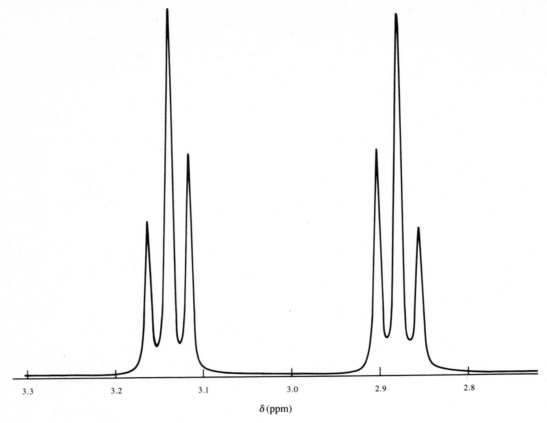

FIGURE 9-8 The partial 300-MHz ^{1}H spectrum of I—CH$_2$—CH$_2$—CO$_2$H.

EXAMPLE 9-14 Explain why multiplets due to heteronuclear coupling exhibit *no* significant departure from first-order predictions.

Solution: Although heteronuclear coupling constants can be quite large (Tables 9-1 and 9-3), the differences in chemical shift are huge. For example, consider the F—C—H system at a field strength of 1.41 T (where the ^{1}H frequency is 60 MHz). Even though $^2J_{HF}$ is 45 Hz, Δv at this field strength is 3,560,000 Hz (3.56 MHz, Table 2-1). With a resulting $\Delta v/J$ ratio of 79,000, first-order behavior can be predicted confidently!

EXAMPLE 9-15 Consider

$$\text{(H}_3\text{C)}_3\text{C} - \overset{\overset{\displaystyle H_c}{|}}{\underset{\underset{\displaystyle Br}{|}}{C}} - \overset{\overset{\displaystyle H_a}{|}}{\underset{\underset{\displaystyle H_b}{|}}{C}} - \overset{\overset{\displaystyle O}{\|}}{C} - \text{C (CH}_3\text{)}_3$$

The 60-MHz ^{1}H spectrum of this compound shows, in addition to the two 9-hydrogen singlets for the (CH$_3$)$_3$C— groups, a set of signals for H$_a$, H$_b$, and H$_c$. Their shifts and their coupling constants are

H	v from TMS (Hz)	J (Hz)	
a	163	ab	−18
b	195	ac	2
c	263	bc	10

Now here are the questions:

(a) Why are H_a and H_b *non*equivalent?

(b) Calculate $\Delta v/J$ for each coupling interaction. What type of spin system do these nuclei constitute?

(c) Using graph paper, draw the predicted *first-order* spectrum for the three hydrogens.

(d) By successively "turning on" the slanting due to a−b, a−c, then b−c coupling, predict the final appearance of the spectrum, including all appropriate multiplet slanting.

(e) What do the magnitudes of $^3J_{ac}$ and $^3J_{bc}$ indicate about the molecule's preferred conformation?

Solution:

(a) H_a and H_b are rendered *diastereotopic* (and hence nonequivalent) by the *neighboring* asymmetric center (the carbon bearing H_c). (Section 4-3).

(b)
$$\left(\frac{\Delta v}{J}\right)_{ab} = \frac{(195 - 163)\ \text{Hz}}{18\ \text{Hz}} = 1.8 \qquad \text{strongly coupled}$$

$$\left(\frac{\Delta v}{J}\right)_{ac} = \frac{(263 - 163)\ \text{Hz}}{2\ \text{Hz}} = 50 \qquad \text{weakly coupled}$$

$$\left(\frac{\Delta v}{J}\right)_{bc} = \frac{(263 - 195)\ \text{Hz}}{10\ \text{Hz}} = 6.8 \qquad \text{moderately coupled}$$

Thus, we could designate this system either ABM or ABX.

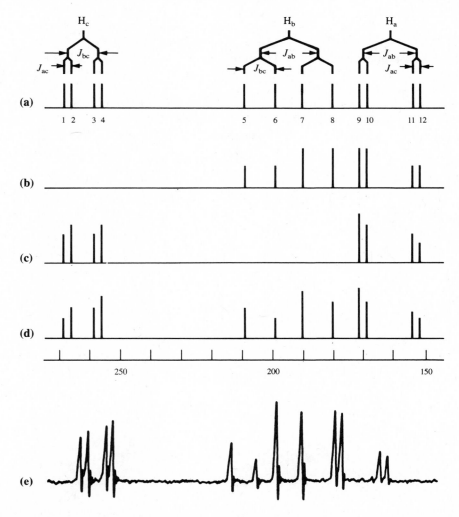

FIGURE 9-9 Predicted (**a–d**) and observed (**e**) ^1H spectra (partial) of $(CH_3)_3C-\overset{\overset{\displaystyle O}{\|}}{C}-CH_2-\overset{\overset{\displaystyle Br}{|}}{CH}-C(CH_3)_3$.

(c) Each of the hydrogens would give rise to a doublet of doublets. The position of each of the twelve equally intense lines can be predicted from the values of ν and J, as in Example 8-11. $|J|$ indicates absolute value.

Line		Position (Hz)
H_c	1	$\nu_c + (|J_{bc}| + |J_{ac}|)/2 = 269$
	2	$\nu_c + (|J_{bc}| - |J_{ac}|)/2 = 267$
	3	$\nu_c - (|J_{bc}| - |J_{ac}|)/2 = 259$
	4	$\nu_c - (|J_{bc}| + |J_{ac}|)/2 = 257$
H_b	5	209
	6	199
	7	191
	8	181
H_a	9	173
	10	171
	11	155
	12	153

These lines are shown in Figure 9-9a.

(d) Figure 9-9b shows the slanting effects due to strong a–b coupling: the *inner* lines of the two coupled multiplets (lines 7, 8, 9, 10) increase in intensity at the expense of the outer lines (5, 6, 11, 12). Figure 9-9c adds slight slanting due to the weak a–c coupling: inner lines 2, 4, 9, and 11 increase at the expense of outer lines 1, 3, 10, and 12. Finally, Figure 9-9d adds the effects of moderate b–c coupling: inner lines 3, 4, 5, and 7 increase, outer lines 1, 2, 6, and 8 decrease. Note how well this qualitative approach predicts the relative intensities observed in the actual spectrum, Figure 9-9e.

(e) Comparing the two ^3J values with the Karplus relation [Figure 9-6 or Eq. (9-3)], we predict that $^3J_{ac}$ (2 Hz) corresponds to a dihedral angle of around 90°, while $^3J_{bc}$ (10 Hz) could correspond to either 20° or 150°. By drawing a structure carefully, we can see that only the 150° angle is possible:

side view   end view

Having completed Example 9-15, you can no doubt begin to appreciate just how complicated a spectrum can become. What recourse do we have when a spectrum is so complex that first-order analysis becomes virtually impossible?

9-10. Calculated Predictions of Spectral Appearance

Quantum mechanics once again comes to our rescue! If we know (or can guess) the values of the chemical shifts and coupling constants for a given structure, it is possible to calculate the *exact* position and intensity of every line in its nmr spectra. This requires the simultaneous solution of all the quantum-mechanical wave equations that describe the spin system. Such calculations normally require so much number crunching that they must be performed by a computer. The most widely used program for this purpose is LAOCOON III,[4] and versions are available for both mainframe and minicomputers.

But suppose you have a very complicated spectrum and only a crude idea of the chemical shifts and coupling constants. (Certainly, after reading chapters 6 through 9, you should have *at least* a crude idea!) No problem! You simply enter your crude guesses, along with the measured positions of each line in the actual spectrum, into the computer. Then the program iteratively fits calculated with observed line positions until it finds a best (least squares) fit. It then lists the best values of all chemical shifts and coupling constants.

The results of such calculations are very impressive, often giving calculated spectra that are superimposable on the actual spectra. For example, Figure 9-10 shows both the calculated and observed ^{1}H spectra of the compound described in Example 9-15. Figure 9-11 shows the computer-generated

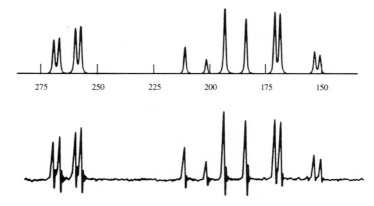

FIGURE 9-10 Computer-generated (**a**) and observed (**b**) 60-MHz ^{1}H spectra of $(CH_3)_3C{-}C{-}CH_2{-}CH{-}C(CH_3)_3$.

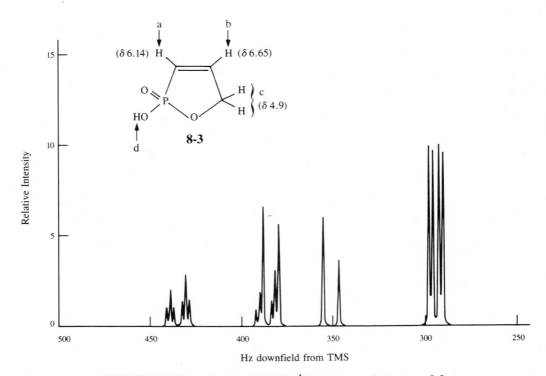

FIGURE 9-11 The calculated 60-MHz ^{1}H spectrum of structure **8-3**.

60-MHz ^{1}H spectrum of structure **8-3** resulting from input of these spectral data:

H	v (Hz)*	J (Hz)	
a	368	P–H$_a$ 33	H$_a$–H$_b$ 8.5
b	399	P–H$_b$ 49	H$_a$–H$_c$ 0
c	294	P–H$_c$ 5.5	H$_b$–H$_c$ 2
d	off scale		

* Downfield from TMS.

All couplings to H$_d$ are erased by exchange processes. The chemical shift of the phosphorus (which would not be part of the ^{1}H spectrum, remember?) can be set to any off-scale value. However, you must enter some value for v_P; otherwise the program will not calculate P–H couplings.

9-11. The AX → AB → A$_2$ Continuum

Although a detailed examination of the calculated spectra for all common spin systems is beyond the scope of this book, there is a compilation of such data for the interested reader.[5] Nonetheless, it is instructive to examine the results in the simplest coupled system, that of two spins.

Let's consider a molecule with two coupled nuclei (A and B) of the same type (e.g., ^{1}H). There are three independent variables that describe the system completely: the chemical shifts of A and B (v_A and v_B, expressed in Hz downfield from TMS), and their homonuclear coupling constant J. The exact appearance of the nmr spectrum for this system, that is, the position and intensity of each line, can be calculated from the values of these three variables alone. The general solution for the two-spin system is

Line	Position*	Relative intensity
1	$\bar{v} + C + \dfrac{J}{2}$	$1 - \dfrac{J}{2C}$
2	$\bar{v} + C - \dfrac{J}{2}$	$1 + \dfrac{J}{2C}$
3	$\bar{v} - C + \dfrac{J}{2}$	$1 + \dfrac{J}{2C}$
4	$\bar{v} - C - \dfrac{J}{2}$	$1 - \dfrac{J}{2C}$

* Hz downfield from TMS

where

$$\bar{v} = \frac{v_A + v_B}{2}, \qquad C = \frac{\sqrt{(\Delta v)^2 + J^2}}{2}, \qquad \text{and} \qquad \Delta v = v_A - v_B$$

EXAMPLE 9-16 Consider two extremes of the two-spin system. Predict the appearance of the spectrum (**a**) when $\Delta v \gg J$, and (**b**) when $\Delta v = 0$.

Solution:

(**a**) Because $\Delta v \gg J$, $\Delta v/J$ is very large. This is the weakly coupled limit, an example of an AX spectrum. In this case

$$C = \frac{\sqrt{(\Delta v)^2 + J^2}}{2} \approx \frac{\sqrt{(\Delta v)^2}}{2} = \frac{\Delta v}{2}$$

and

$$\frac{J}{2C} = \frac{J}{\Delta v} \approx 0$$

Substituting these "values" for C and $J/2C$ gives a spectrum described by

Line	Position	Relative intensity
1	$v + \dfrac{\Delta v}{2} + \dfrac{J}{2}$	1
2	$v + \dfrac{\Delta v}{2} - \dfrac{J}{2}$	1
3	$v - \dfrac{\Delta v}{2} + \dfrac{J}{2}$	1
4	$v - \dfrac{\Delta v}{2} - \dfrac{J}{2}$	1

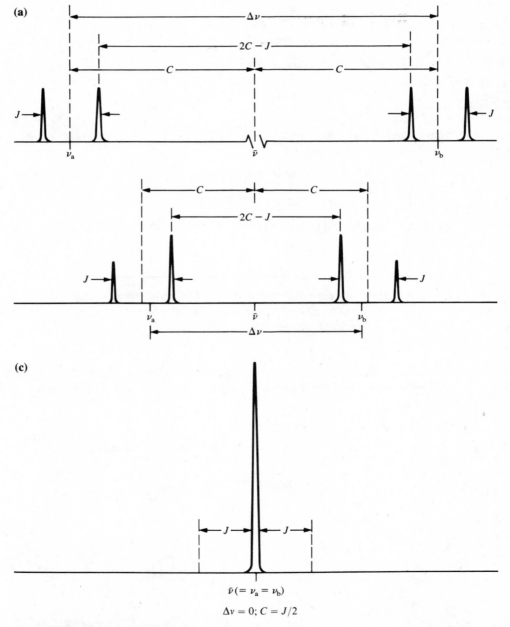

FIGURE 9-12 The AB spectrum as a function of $\Delta v/J$. (a) $\Delta v/J > 10$; (b) $0 < \Delta v/J < 10$; (c) $\Delta v/J = 0$.

This result, shown in Figure 9-12a, is exactly the two doublets we would have expected from the first-order $(n + 1)$ rule.

(b) If $\Delta v = 0$, nuclei A and B have the same chemical shift and so are chemically equivalent, making this an A_2 spin system. In this case,

$$C = \frac{\sqrt{(J^2)}}{2} = \frac{J}{2}$$

and

$$\frac{J}{2C} = 1$$

Notice how these values affect the position and especially the *intensity* of each line:

Line	Position	Relative intensity
1	$\bar{v} + J$	0
2	$\bar{v}$	2
3	$\bar{v}$	2
4	$\bar{v} - J$	0

This result, shown in Figure 9-12c, has lines 2 and 3 superimposed at $\bar{v}$ ($= v_A = v_B$), while outer lines 1 and 4 have *zero* intensity. Now you see why coupling between equivalent nuclei, though it *does* occur, is not evident in the spectrum (Section 8-4)!

A graphical representation of the general solution for the AB system is shown in Figure 9-12b. The most significant features to remember are

1. The AB spectrum is symmetrical around its midpoint $(\bar{v})$, with the inner lines larger and the outer lines smaller.
2. The doublets are centered *not* at v_A and v_B but rather at $\bar{v} \pm C$.

The latter feature is what makes the measurement of exact chemical shifts difficult in spectra that show second-order effects.

EXAMPLE 9-17 Suppose there were an AB spectrum in which the spacing between lines 2 and 3 accidentally equaled J. **(a)** What would be the position and intensity of each line? **(b)** What would be the value of $\Delta v/J$? **(c)** How would this spectrum compare with a first-order quartet?

Solution:

(a) From either Figure 9-12b or the general equations of the two-spin system, we can see that the separation between lines 2 and 3 is $2C - J$. But in this case we've set this separation equal to J. Therefore,

$$2C - J = J \quad \text{or} \quad C = J$$

Using this relation, we can calculate the position and intensity of each line:

Line	Position	Relative intensity
1	$\bar{v} + \dfrac{3J}{2}$	$\dfrac{1}{2}$
2	$\bar{v} + \dfrac{J}{2}$	$\dfrac{3}{2}$
3	$\bar{v} - \dfrac{J}{2}$	$\dfrac{3}{2}$
4	$\bar{v} - \dfrac{3J}{2}$	$\dfrac{1}{2}$

(b) From part (a), we know that $C = J$. Therefore,

$$J = C = \frac{\sqrt{(\Delta v)^2 + J^2}}{2}$$

$$2J = \sqrt{(\Delta v)^2 + J^2}$$

$$4J^2 = (\Delta v)^2 + J^2$$

$$3J^2 = (\Delta v)^2$$

$$\Delta v/J = \sqrt{3}$$

(c) The result in (a), shown in Figure 9-13, is identical to a first-order quartet with the intensity ratios predicted by Pascal's triangle. Are you wondering how to tell whether a quartet is a "real" quartet or an "AB quartet"? Well, there are two ways. If it's a "real" quartet, there are three other equivalent nuclei somewhere in the molecule, responsible for the coupling; find them! If you can't find them (and remember, they may be heteronuclei), try generating another spectrum at a higher field strength. A "real" quartet will be unaffected, whereas in an "AB quartet" the spacing (in Hz) between lines 2 and 3 will increase while the spacing between lines 1 and 2 and between lines 3 and 4 stays constant as Δv increases (Section 9-9).

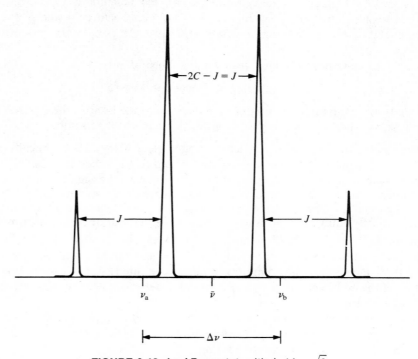

FIGURE 9-13 An AB quartet, with $\Delta v/J = \sqrt{3}$.

EXAMPLE 9-18 Values of how many quantities are needed to calculate the appearance of an AA′BB′ spectrum?

Solution: Two chemical shifts, for A (= A′) and B (= B′), and four coupling constants, A–B, A′–B′, A′–B, and A–B′. Often, fewer coupling constants are needed, either because $J_{AB} = J_{A'B'}$ and $J_{A'B} = J_{AB'}$, or because $J_{AB'} = J_{A'B} = 0$.

Now that we've spent two chapters describing spin–spin coupling, let's move on. Ahead, in Chapter 11, we'll discuss ways to get rid of the effects of coupling!

REFERENCES

1. E. D. Becker, *High Resolution NMR*, 2nd ed. Academic Press, New York, 1980.

2. P. Lazlo and P. J. Stang, *Organic Spectroscopy*. Harper and Row, New York, 1971.

3. R. M. Silverstein, G. C. Bassler, and T. C. Morrill, *Spectrometric Identification of Organic Compounds*, 4th ed. Wiley, New York, 1981.

4. Written by S. Castellano and A. A. Bothner-By, and available from the Quantum Chemistry Program Exchange at Indiana University, Bloomington, IN 47401.

5. K. B. Wiberg and B. J. Nist, *Identification of NMR Spectra*. Benjamin, New York, 1962.

SUMMARY

1. Nuclear spin coupling constants (J values) are either positive or negative. If the value of J is *positive*, the antiparallel arrangement of nuclear spins of the coupled nuclei is lower in energy than the parallel arrangement; if the value of J is *negative*, the parallel arrangement is lower in energy than the antiparallel arrangement.

2. In general, the signs of H–H, C–H and C–C coupling constants are a function of the number of intervening bonds: *positive* if the number of bonds is *odd*, *negative* if the number of bonds is *even*.

3. Normally, the *sign* of a coupling constant has no effect on the appearance of the nmr spectrum.

4. The *magnitude* of coupling constants is dependent on many factors, including

   (a) The number of intervening bonds: J normally decreases as the number of bonds increases.

   (b) The product of the magnetogyric ratios of the coupled nuclei: $J \propto \gamma_1 \gamma_2$.

   (c) The fraction of s-character (f_s) of the hybrid orbitals connecting the nuclei: $J \propto f_{s_1} f_{s_2}$.

   (d) For geminal (two-bond) coupling constants, the internuclear angle: ^2J increases (becomes more positive) as the angle increases.

   (e) For vicinal (three-bond) coupling constants, the dihedral angle θ:

   $$^3J_{HH} = 7 - \cos\theta + 5\cos^2\theta \tag{9-3}$$

   (f) For long-range coupling constants (over more than three bonds), the number of intervening multiple bonds and the geometric relationship between the orbitals involved.

5. Two or more nuclei are said to be *magnetically* equivalent if they are chemically equivalent (i.e., possess the same chemical shift) *and* are equally coupled to any other nucleus.

6. Spin systems (collections of interacting nuclei) are often labeled by assigning each set of magnetically equivalent nuclei a letter from the alphabet. Nuclei that are close (but not identical) in chemical shift are given letters that are close in the alphabet (e.g., AB). Two nuclei that are chemically equivalent but *not* magnetically equivalent are assigned the same letter, but one letter is primed (e.g., AA').

7. As the ratio of $\Delta\delta$ (the difference in chemical shift between two coupled nuclei) to J DECREASES, the relative intensities of the lines in a multiplet deviate farther from first-order (Pascal triangle) ratios. Inner lines (those facing the coupled multiplet) increase in intensity, while outer lines lose intensity. This slanting of the multiplets is referred to a second-order effect. At very small values of $\Delta\delta/J$ extra lines may begin to appear in the multiplets as well.

8. Computer programs exist that can calculate the exact position and intensity of each line in an nmr spectrum for which all δ and J values are known. Such programs can also extract exact δ and J values from a spectrum by iterative curve fitting.

10 ESR and CIDNP

10-1. An Electron in a Magnetic Field

From Section 2-1B you may recall that an electron, like a proton, has magnetic properties ($I = \frac{1}{2}$) and that it can adopt either of two spin orientations in a magnetic field (Figure 2-1). Moreover, as with protons, electrons in these two spin orientations have different energies but precess at the same frequency. It should, therefore, come as no surprise that it is possible to induce transitions between these *electronic* spin energy levels by subjecting the electrons to electromagnetic radiation of a frequency equal to their precessional frequency. In fact, this is the basis of **electron spin resonance** (esr) **spectroscopy**, also known as **electron paramagnetic resonance** (epr) **spectroscopy**.

10-2. Free Radicals

There is a fly in the ointment, though. Esr spectroscopy can detect only species that possess one or more *unpaired* (Sections 2-1B and 9-2) electrons. Such species are said to be *paramagnetic*; their magnetic moments align *with* an applied magnetic field and they are weakly attracted into the field. Virtually all *stable* organic molecules have their electrons arranged in pairs and are therefore *diamagnetic;* they are weakly repelled from a magnetic field. Esr is of no use for studying diamagnetic compounds. The few organic molecules that *are* paramagnetic, called **free radicals** (not a reflection of their politics), are usually extremely reactive species because they have an unpaired electron. They are often formed as short-lived **intermediates** during chemical reactions.

   Free radicals are usually formed as the result of cleavage of a chemical bond which, you'll recall, is nothing more than a *pair* of electrons (Section 9-2). This cleavage requires an amount of energy called, logically enough, the **bond dissociation energy**. Once formed, these free radicals usually exist only until they find a way to pair up their odd electron. The most obvious way to accomplish this is by encountering another free radical and re-forming the bond, a process called **radical recombination**. An unpaired electron in a structure is shown either as a dot or a small arrow:

10-3. The Landé Factor, *g*

The precessional frequency of an electron is directly proportional to the applied magnetic field strength and can be expressed by the equation

ELECTRON PRECESSIONAL FREQUENCY $v_{prec} = (\text{constant})gB_0$ **(10-1)**

Notice how similar this relationship is to the one that describes *nuclear* precession [Eq. (2-6)], except that the magnetogyric ratio (γ) in the nuclear equation is replaced by the *g* value of the electron, known as its **Landé factor**. Just as the value of γ determines the precessional frequency of a nucleus in a magnetic field, the value of *g* determines the precessional frequency of an unpaired electron. Moreover, just as changes in molecular environment and shielding cause nuclei to precess at *slightly* different frequencies (Section 6-1), the exact *g* value of the unpaired electron in a free radical depends on the structure of that radical. So, we will adopt the *g* value as our position parameter ("chemical shift," if you will) in esr spectroscopy. Table 10-1 lists the *g* values of several different free radicals, as well as the value for a "free" electron. At first glance, it might seem unwise to use the *g* value as the position parameter, since all the values appear to be so similar. But, as Example 10-1 shows, these "small" differences in *g* actually correspond to large (and easily measured) differences in frequency.

TABLE 10-1 *g* Values of Several Free Radicals[a]

Structure	*g*
$H_2C{=}\overset{\cdot}{C}H$	2.002 20
free electron	2.002 32
$H_2C{=}CH{-}\overset{\cdot}{C}H_2$	2.002 54
$H_3C\cdot$	2.002 55
$H_3C{-}\overset{\cdot}{C}H_2$	2.002 60
(cyclohexadienyl radical)	2.002 76
$\bar{:}O{-}\langle\bigcirc\rangle{-}O\cdot$	2.003 59
$R{-}C\big({=}O\big)O\cdot$	2.0058
$Cl_3C\cdot$	2.0091
$R{-}O{-}O\cdot$	2.0155

[a] Data from reference 1 at the end of this chapter.

EXAMPLE 10-1 A free electron precesses at a frequency of 9.500 GHz (1 gigahertz = 1000 MHz = 10^9 Hz) at a field strength of 0.3400 T. **(a)** Evaluate the constant in Eq. (10-1). **(b)** Calculate the *difference* between the precessional frequencies of $Cl_3C\cdot$ and $R{-}CO_2\cdot$ at 0.3400 T. **(c)** What would be the precessional frequency of a proton at 0.34 T?

Solution:

(a) Solving Eq. (10-1) for the constant, we find

$$(\text{constant}) = \frac{v_{\text{prec}}}{gB_0} = \frac{9.500 \text{ GHz}}{(2.002\,32)(0.3400 \text{ T})} = 13.95 \text{ GHz T}^{-1}$$

(b) The difference between two precessional frequencies can be most easily calculated from another version of Eq. (10-1):

$$\Delta v = v_1 - v_2 = (\text{constant})(g_1 - g_2)(B_0)$$

$$= (13.95 \text{ GHz T}^{-1})(2.0091 - 2.0058)(0.3400 \text{ T})$$

$$= 0.016 \text{ GHz} = 16 \text{ MHz} = 1.6 \times 10^7 \text{ Hz!}$$

Remember that the larger the g value, the higher the precessional frequency of the electron at a given field strength.

(c) Use Eq. (2-6), with $\gamma = 267.5 \times 10^6 \text{ rad T}^{-1}\text{s}^{-1}$ (Table 2-1):

$$v_{\text{prec}} = \frac{\gamma B_0}{2\pi} = \frac{(267.5 \times 10^6 \text{ rad T}^{-1}\text{s}^{-1})(0.34 \text{ T})}{2\pi \text{ rad}}$$

$$= 14.5 \times 10^6 \text{ Hz} = 14.5 \text{ MHz}$$

Compare this value to the electron's precessional frequency of 9500 MHz!

10-4. Electron Spin Energy Levels

As you can see from Example 10-1c, an electron precesses 655 times as fast as a proton at the same field strength. Therefore, even small differences in g correspond to large differences in v_{prec}. This fact has several important consequences. For example, because the frequencies and frequency differences are so large in esr, small inhomogeneities and fluctuations in magnetic field strength do not affect signal position significantly, as they do in nmr (Section 3-1A). Therefore, neither a lock substance (Section 3-1A) nor an internal reference compou d (Section 5-1A) need be added to the sample. However, an external paramagnetic reference standard (e.g., Cr^{3+} in MgO) *is* used to calibrate the instrument. To understand some of the other consequences of the electron's higher precessional frequency, we need to review the characteristics of spin energy levels.

Figure 10-1 shows how the two energy levels of an electron vary with field strength. The only differences between this figure and the one for proton spin states (Figure 2-3a) is that the orientations are reversed (the one *antiparallel* to the applied field is lower in energy) and the energy gap at a given field strength is 655 times that for protons.

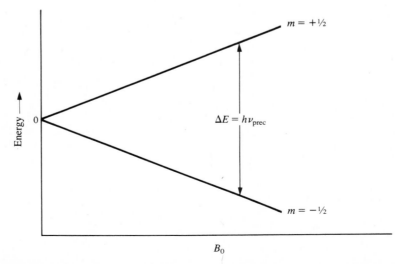

FIGURE 10-1 Variation of spin state energy as a function of magnetic field strength.

EXAMPLE 10-2 **(a)** What is the energy gap (in joules) between the two spin states of an electron at 0.34 T? **(b)** What is the equilibrium population ratio of these states at room temperature?

Solution:

(a) Recall from Section 2-3A that the energy gap between two spin states is proportional to the precession frequency, which must be matched by the irradiation frequency. From our value of ν_{prec} in Example 10-1a,

$$\Delta E = h\nu_{prec} = (6.63 \times 10^{-34}\,J\,s)(9.5 \times 10^9\,s^{-1}) = 6.30 \times 10^{-24}\,J$$

(b) Do you remember the Boltzmann distribution [Eq. (2-8)]?

$$\frac{P_{m=1/2}}{P_{m=-1/2}} = e^{-\Delta E/kT} = e^{-(6.30 \times 10^{-24}\,J)/(1.38 \times 10^{-23}\,JK^{-1})(298\,K)} = 0.998486$$

This ratio indicates that at equilibrium 50.0379% of the electrons are in the lower ($m = -\frac{1}{2}$) state, while 49.9621% are in the upper state. Or, to put it another way, of one million electrons, there are 758 more in the lower state than in the upper state. Although this may seem like a very small difference, it is much greater than the difference in proton spin state populations, a difference of only 8 nuclei per million at 2.35 T (Example 2-8)!

The fact that the population difference between spin states is greater for electrons than for nuclei means that esr spectroscopy is much more sensitive than nmr. Thus, while modern PFT nmr methods (Section 3-3) still require sample concentrations of at least 0.01 *M* (moles per liter), esr signals can be detected from radicals in as low a concentration as 10^{-8} *M* at room temperature (in a sample volume of 0.2 mL). Moreover, because esr circumvents the problems we encounter in nmr with solid-phase samples (Section 12-3), esr spectra can routinely detect the presence of free radicals in solid samples or in glassy matrices.

As with nmr (Section 3-2), an esr spectrometer can in principle use either the frequency sweep or the field sweep mode. In practice, the latter technique is easier to use, because varying an rf frequency of 9.5 GHz smoothly over a large enough range is difficult.

In Chapter 1 we discussed the concept of spectroscopic time scale. Because esr involves frequencies on the order of 10^9 Hz (and a resulting time scale of 10^{-9} s), it takes a much faster "snapshot" of dynamic systems than does ^{1}H nmr. As a result, esr can generate information about chemical processes that are too fast to study by nmr.

There is one other difference you will notice between an esr signal and an nmr signal. All the nmr spectra illustrated so far in this book are displayed in the **absorption mode**. Practical considerations, however, dictate that it is easier to portray an esr signal in the dispersion mode (Figure 10-2). If you have had any differential calculus, you may recognize a dispersion signal as essentially the first derivative (slope) of the corresponding absorption signal.

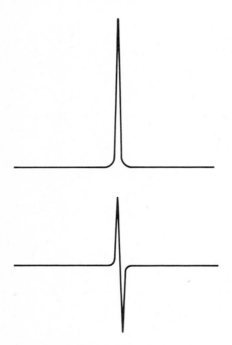

FIGURE 10-2 An absorption signal (top) compared to a dispersion signal (bottom). The position of a dispersion signal is where it crosses the horizontal axis.

10-5. Hyperfine Coupling and the *a* Value

Just as the nmr signal of one nucleus can be split by other neighboring nuclei (Chapters 8 and 9), the esr signal of the unpaired electron in a radical can be split by magnetic nuclei within the radical. This coupling follows *exactly* the same first-order (Pascal's triangle) rule we developed for protons. Thus, the number of lines in an esr multiplet is governed by Eq. (8-3). The only differences are that the separation between the lines of a multiplet in an esr spectrum is called **hyperfine coupling**, or hfc (rather than coupling constant), is given the symbol a (rather than J), and is expressed in the unit of magnetic field strength *gauss* (rather than in Hz).

EXAMPLE 10-3 Predict the appearance of the esr spectrum of $CH_3\cdot$.

Solution: The signal will be centered at a g value of 2.002 55 (Table 10-1), and it will be split into a quartet (intensity ratio 1:3:3:1) by the three neighboring hydrogens.

As with the coupling constant J, the magnitude of a depends on the "proximity" of the unpaired electron and the coupling nuclei. But the word "proximity" is harder to define for an electron than for a nucleus. Actually, the unpaired electron is to some extent "spread out" through the molecule by resonance (Sections 7-4 and 9-6), spending a (not necessarily equal) portion of its time near each nucleus. The fraction of time it spends near a given nucleus is called the **free electron density** (ρ) at that nucleus, and the magnitude of a is directly proportional to ρ. In fact, the value of a is the best experimental measure of how the free electron density is distributed. For example, the magnitude of a for the methyl radical (Example 10-3) is 23.0 G. Similarly, the magnitudes of a for the propyl radical (structure **10-1**) are $a_\alpha = 22.1$ G, $a_\beta = 33.2$ G, and $a_\gamma = 0.4$ G.[1]

$$H_\gamma\!\!-\!\!\underset{\underset{\textstyle H_\gamma}{|}}{\overset{\overset{\textstyle H_\gamma}{|}}{C}}\!\!-\!\!\underset{\underset{\textstyle H_\beta}{|}}{\overset{\overset{\textstyle H_\beta}{|}}{C}}\!\!-\!\!\underset{\underset{\textstyle H_\alpha}{|}}{\overset{\overset{\textstyle H_\alpha}{|}}{C}}\cdot$$

propyl radical

10-1

It may surprise you that a_β is larger than a_α, but this fact is consistent with the way the unpaired electron is distributed throughout the molecule. In particular, another contributing **hyperconjugative resonance structure** for the propyl radical can be written with the electron on H_β:

$$H\!\!-\!\!\underset{\underset{\textstyle H}{|}}{\overset{\overset{\textstyle H}{|}}{C}}\!\!-\!\!\underset{\underset{\textstyle H}{|}}{\overset{\overset{\textstyle H\cdot}{|}}{C}}\!\!=\!\!\underset{\underset{\textstyle H}{|}}{\overset{\overset{\textstyle H}{|}}{C}}$$

No satisfactory structure can be written that places the electron on H_α or H_γ.

EXAMPLE 10-4 How many lines would you expect to find in the esr spectrum of the propyl radical, and where would the signal be centered?

Solution: From Table 10-1 we can guess that the signal should occur near $g = 2.002\,60$. It will be split into a triplet (by the two H_β's) of triplets (by the two H_α's) of quartets (by the three H_γ's), a total of 36 lines! Alternatively, we could have used Eq. (8-3):

$$L = (2 \cdot 2 \cdot \tfrac{1}{2} + 1)(2 \cdot 2 \cdot \tfrac{1}{2} + 1)(2 \cdot 3 \cdot \tfrac{1}{2} + 1) = 36$$

Notice that we can neglect hyperfine coupling to the carbons because of the low natural abundance of ^{13}C (Table 2-1).

EXAMPLE 10-5 An a value of 33.2 G at 0.34 T and 9.5 GHz is equivalent to what value in Hz? Recall that 1 T = 10,000 G.

Solution: We can set up a simple proportion:

$$\frac{33.2 \text{ G}}{3400 \text{ G}} = \frac{x}{9.5 \times 10^9 \text{ Hz}}$$

$$x = 9.28 \times 10^7 \text{ Hz} = 92.8 \text{ MHz}$$

Compare this to typical $^1H\!-\!^1H$ coupling constants of 1–20 Hz!

As with homo- and heteronuclear coupling constants, *a* values also have *signs*. A useful generalization (though not without exceptions) is that the sign of *a* is a function of the number of bonds separating the unpaired electron from the coupling nucleus. If the separation is zero or an even number of bonds, *a* will be positive; if the separation is an odd number of bonds, *a* will be negative. Note that this pattern is exactly the reverse of the pattern we saw for nuclear coupling constants (Section 9-2) and is due to the *negative* charge of the electron.

EXAMPLE 10-6 Predict the *sign* of the *a* value of the methyl radical and of the a_α and a_β values for the propyl radical.

Solution: For $CH_3 \cdot$ (one bond separation), the *a* value is negative (-23.0 G). For the propyl radical, a_α (one bond separation) is negative (-22.1 G) and a_β (two bond separations) is positive ($+33.2$ G).

Although the *sign* of *a* has no effect on the appearance of an esr spectrum (just as the sign of *J* has no effect on an nmr spectrum, Section 9-1), it *does* have an effect in CIDNP (Section 10-7).

10-6. A Typical ESR Spectrum

Because resonance interactions play such an important role in determining the appearance of esr spectra, let's consider one final example of such effects. The compound with structure **10-2**, BHT, is a common antioxidant that owes its chemical properties to the ease with which it forms phenoxy radical, structure **10-3**. This radical exists as a hybrid of five principal resonance structures, **10-3a–e**. From these structures several things are apparent. First, the overall structure of **10-3** has a plane of symmetry relating its left and right halves. Second, the electron spin density is concentrated mainly on the oxygen (in **a**), on the ortho carbons (**b** and **c**), on the para carbon (**d**), and on the methyl hydrogens (**e**).

EXAMPLE 10-7 Predict the appearance of the esr spectrum of phenoxy radical **10-3**.

Solution: The signal should be centered at the position of resonance-stabilized oxygen radicals (a *g* value in the range 2.0036–2.0058, Table 10-1). It will be split into a *quartet* by the three methyl hydrogens, and each line of the quartet will be further split into a *triplet* by the two equivalent ring hydrogens. The *a* value for the former coupling (zero-bond separation) should be larger than that for the latter coupling (two-bond separation). Further splitting by the eighteen equivalent *tert*-butyl hydrogens should be negligible (three-bond separation), and there will be no splitting by oxygen (^{16}O is nonmagnetic, Table 2-1).

$g = 2.004$

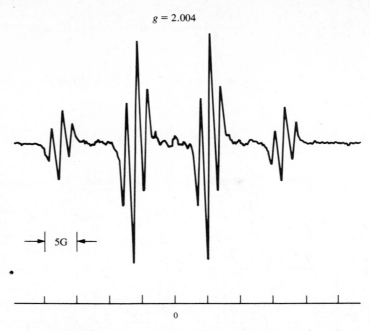

5G

0

FIGURE 10-3 The esr spectrum of phenoxy radical **10-3**.

The actual esr spectrum of **10-3** is shown in Figure 10-3. The quartet of triplets is centered at $g = 2.0040$, with a values of $+11.3$ and $+1.5$ G, in total agreement with our predictions.

10-7. CIDNP: Mysterious Behavior of an NMR Spectrometer

Back in 1967 two groups of chemists, working independently, were using nmr spectroscopy to study certain chemical reactions. They became quite agitated when the signals portrayed on their spectra seemed to make no sense. Some of the signals were much more intense than they should have been, while other signals were *negative* (upside down). And in some cases multiplets were composed of both positive and negative signals! Only after the nmr repairman assured them that their instruments were in perfect working order did it become apparent that some new effect had been discovered. This effect has come to be known as **chemically induced dynamic nuclear polarization**, or CIDNP.

10-8. The Net Effect

In order to understand these strange observations, we need to review the importance of nuclear spin state populations on nmr signal intensity. As first discussed in Section 2-3 (and again in Section 10-4), the intensity of an nmr signal depends on the ability of the spin system to absorb photons of electromagnetic radiation. There are two requirements for this to happen: The frequency of the radiation must match the precessional frequency, *and* there must be more nuclei in the lower energy state than in the upper state. Because at equilibrium this *difference* in spin state populations is very small (about 8 parts per million, Example 2-8), any factor that affects the relative populations of the two states can have a profound effect on signal intensity. For example, increasing relaxation rates (Section 2-3), increasing magnetic field strength, or decreasing the temperature of the sample all serve to increase the equilibrium population of the lower state and increase signal intensity. Conversely, if relaxation is slow, the system can become saturated (equal populations in both states, Section 2-3) so that no net absorption is possible. It turns out that chemical reactions of certain types can form products whose nuclei have *non*equilibrium spin state populations, and these reactions give rise to the dramatic CIDNP effects. We describe a spin system with nonequilibrium spin state populations as *polarized*.

Suppose, for example, a reaction product is formed in which the nuclei have a polarized spin state distribution, with more than the usual (equilibrium) number in the lower state. This would make the absorption process more favorable, and the resulting signal intensity would be greater than normal. This situation is referred to as enhanced absorption, or an *A* **net effect**. On the other hand, suppose for some strange reason the spin state distribution were polarized in the opposite way, with more nuclei in the

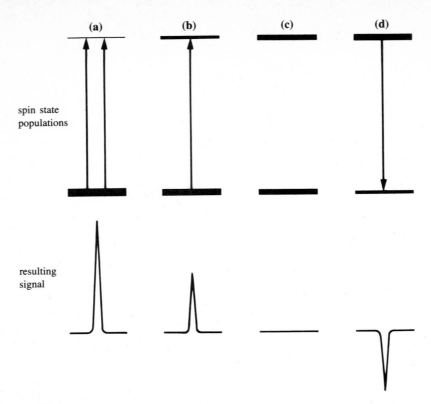

FIGURE 10-4 Effect of spin state populations on nmr signal intensity. Thickness of spin state line indicates relative population. **(a)** Large excess in lower state, giving enhanced absorption; **(b)** normal slight excess in lower state, giving normal signal; **(c)** saturation, giving zero signal; **(d)** excess in the upper state, leading to emission.

upper state. Now, not only is absorption disfavored, but net *emission* of radiation (at the precessional frequency) takes place as the molecules attempt to reestablish equilibrium. This emission of radiation, called an ***E* net effect**, results in a negative nmr signal. These effects are depicted in Figure 10-4.

10-9. The Multiplet Effect

As mentioned in Section 10-7, the most baffling aspect of CIDNP spectra was the occurrence of both positive and negative signals in the same multiplet. Sometimes the left half of the multiplet was positive and the right half negative (an *A/E* **multiplet effect**), while sometimes the reverse was observed (an *E/A* **multiplet effect**). These are shown in Figure 10-5.

   As with the net *E* and *A* effects, the multiplet effects can also be rationalized on the basis of nonequilibrium spin state populations. Recall the AB spin system we discussed in Section 9-1. Take a look at Figure 10-6b (page142), a replica of Figure 9-2a. Suppose a chemical reaction generated a product with an AB spin system and $J > 0$ and in which the polarization favored the *parallel* spin states (Figure 10-6a). Lines 1 and 2 would exhibit an *A* effect, while lines 3 and 4 would exhibit an *E* effect, resulting in *E/A* multiplet effects in both doublets. Alternatively, if the *antiparallel* states were favored (Figure 10-6c), *A/E* multiplet effects would be observed in both doublets.

EXAMPLE 10-8 How would a *negative* value of J affect the type of multiplet effects in the CIDNP spectra of an AB system?

Solution: You may remember that changing the sign of J changes the ordering of the lines of each multiplet (Figure 9-2b). If the system were now polarized in favor of the *parallel* states, we would observe *A/E* multiplet effects, while *antiparallel* polarization would result in *E/A* multiplet effects. Thus,

• The sign of J is one factor that governs whether the multiplet effect is *A/E* or *E/A*.

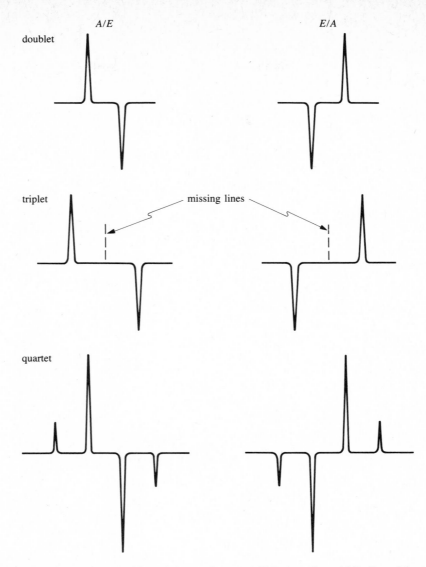

FIGURE 10-5 Multiplet effects in a doublet, triplet, and quartet. Note how the middle line of the triplet vanishes.

10-10. The Radical-Pair Theory of the Net Effect

Initial explanations of CIDNP were based solely on Overhauser effects between an unpaired electron and a nucleus (Example 11-4), but these early theories failed to account for several important aspects of CIDNP, such as multiplet effects. Nearly two years passed before a satisfactory theory explaining CIDNP emerged. It is now believed that chemical reactions involving *pairs* of interacting free radicals (Section 10-2) are responsible for all CIDNP effects.

When the unpaired electrons of two radicals interact within a magnetic field, there are four possible combinations of their precessing spins. One of these has the spins oriented opposite each other (antiparallel), with their precession exactly 180° out of phase. This arrangement, shown in Figure 10-7a, is called the **singlet state**, and it has *zero* magnetic moment. The other three arrangements, referred to collectively as the **triplet state**, have the electrons either parallel "up" ($s = +1$), parallel "down" ($s = -1$), or antiparallel ($s = 0$), as shown in Figure 10-7b, but their precession is *in phase* in all three. Each of the components of the triplet state has a *nonzero* magnetic moment, as shown by the dashed arrows in Figure 10-7b. As you can see, the only difference between the singlet state and the $s = 0$ triplet is the *phase* relationship of the electrons' precessional motion. Now, if both electrons are equivalent (having the same g value, Section 10-3), their phase relationship will remain constant. That is, a singlet will remain a singlet, and a triplet will remain a triplet. *But*, if the two electrons precess at *different* frequencies (because their g values are different), their relative phasing will change with time, causing the radical pair to change from the $s = 0$ triplet state to the singlet state and back again. This process is called **singlet–triplet**

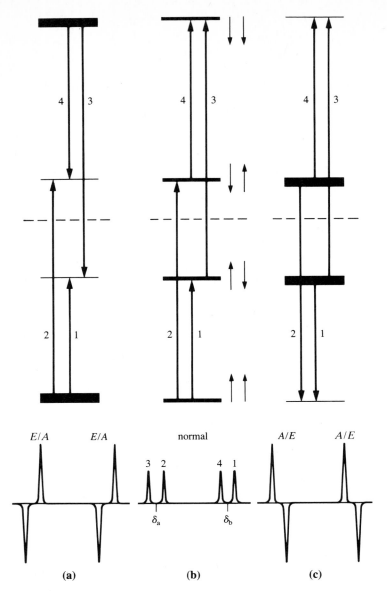

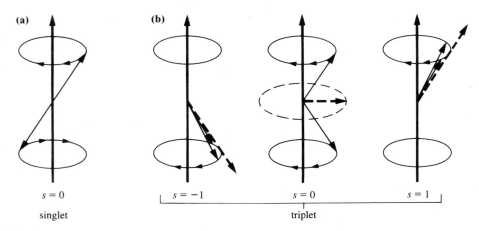

FIGURE 10-6 The way in which spin state populations give rise to the multiplet effect.

FIGURE 10-7 The **(a)** singlet and **(b)** triplet electronic states. Vertical arrow indicates direction of the external magnetic field. Dashed arrows represent the net magnetic vector of the electrons.

mixing. Therefore, any factor (such as hyperfine coupling to a nearby nucleus) that affects the precessional frequency (by splitting it into several values) will also influence the rate of dephasing and rephasing.

When two radicals are formed in solution as the result of bond cleavage, they remain close together for a short time, held in place by a "cage" of solvent molecules. In order for the radicals to recombine (re-forming a bond, Section 10-2), the two electrons must be *paired*, i.e., in the singlet state. If they are *not* in the singlet state, the radicals must stay together long enough for the triplet to rephase into the singlet. If this triplet–singlet mixing takes too long, the radicals may escape the solvent cage *before* rephasing occurs and go on to other types of reactions, forming so-called **escape products**. Thus, recombination products will be formed from radical pairs with nuclear spins that favor the singlet electronic state, while escape products will be formed from radical pairs whose nuclear spins favor the triplet electronic state. This effect can be called *nuclear spin sorting*.

The direction of polarization, and hence the type of net CIDNP effect (*A* or *E*), can be predicted if you know the signs of four variables:

(1) μ, the original spin state of the radical pair at its "birth" ($+$ for triplet, $-$ for singlet);
(2) ε, the type of reaction leading to the observed product ($+$ for cage recombination, $-$ for escape);
(3) $\Delta g = g_1 - g_2$, the difference in g values of the two radicals, where g_1 refers to the radical with the nucleus being observed; and
(4) the sign of a, the hyperfine coupling between the electron and the nucleus being observed.

The type of net effect, given the symbol Γ_n, is determined by the sign of the product of the four variables listed above:

NET EFFECT
$$\Gamma_n = (\mu)(\varepsilon)(\Delta g)(a) \tag{10-2}$$

• A positive Γ_n corresponds to an *A* net effect; a negative Γ_n corresponds to an *E* net effect.

The magnitude of the effect is highly variable and depends on several factors. However, the maximum *A* net effect can result in an intensity enhancement by a factor of several hundred (see Section 11-3).

EXAMPLE 10-9 What type of net effect can be observed between a pair of identical radicals?

Solution: None! Because $\Delta g = 0$, there will be no dephasing or rephasing of the spins. Moral: To generate a *net* effect, two radicals must have *different* g values.

EXAMPLE 10-10 When the peroxide with structure **10-4** is heated in the presence of iodine, four major products are formed: carbon dioxide and compounds **10-5**, **10-6**, and **10-7**.

$$Cl_3C-\overset{\overset{\displaystyle O}{\|}}{C}-O-O-\overset{\overset{\displaystyle O}{\|}}{C}-CH_3 \xrightarrow{I_2} 2\,CO_2 + I-CH_3 + Cl_3C-I + Cl_3C-CH_3$$

10-4   **10-5**   **10-6**   **10-7**

When the reaction is monitored by ^1H nmr, the signal for **10-5** (δ 2.1 ppm) exhibits a strong *A* net effect, while the signal for **10-7** (δ 2.7 ppm) shows a strong *E* net effect. **(a)** Rationalize these CIDNP results. **(b)** Why doesn't **10-6** show a net effect?

Solution:

(a) Heating **10-4** causes the O—O bond to break, leading ultimately to a pair of radicals ($\cdot CH_3$ and $\cdot CCl_3$) and the expulsion of two molecules of CO_2. The horizontal bar in the following equation is there to remind us that the particles are formed within a solvent cage:

$$\text{10-4} \longrightarrow \overline{H_3C\cdot \quad \cdot CCl_3} + 2\,CO_2$$

recombination ↙ I_2 ↘ escape

$$H_3C-CCl_3 \qquad\qquad H_3C-I + I-CCl_3$$

10-7   **10-5**   **10-6**

Either the radical pair can undergo cage recombination to form **10-7** or the radicals can escape and react with I_2 to form **10-5** and **10-6**.

The signal for **10-7** shows net emission, so Γ_n must be negative. What does this tell us about μ, the original spin state of the radical pair? We know the g values for both radicals (Table 10-1), and therefore $\Delta g = 2.0025 - 2.0091$, which is negative. The value of a for $\cdot CH_3$ is negative (Section 10-5), and ε is positive for a recombination product. Thus, expressing each variable in Eq. (10-2) by its sign, we have

$$\Gamma_n = (-) = \mu\,(+)(-)(-)$$

This equation can be true only if μ is also negative, indicating that the radical pair was initially formed in the singlet state. This is eminently reasonable when you consider that precursor **10-4** had all its electrons paired. This result is consistent with the notion that thermal fragmentation of a chemical bond preserves the spin state of the electrons (normally singlet). Many photochemical fragmentations, on the other hand, proceed by way of electronically excited states and produce triplet radical pairs.

Using the negative value of μ deduced above, let's predict the type of net effect for escape product **10-5**. Everything is the same as before, except that ε is now negative (for an escape product). Thus,

$$\Gamma_n = (-)(-)(-)(-) = (+)$$

So, enhanced absorption is predicted, in agreement with experimental observation.

(b) This is a trick question! Since $ICCl_3$ has no hydrogens, it produces no ^1H nmr signal! However, it might show CIDNP effects in its ^{13}C spectrum (see Example 10-12).

From Example 10-10 you can safely infer that the greatest value of CIDNP is its ability to confirm the involvement of radical pairs in reactions and to determine the spin state of a radical pair at its "birth."

10-11. The Radical-Pair Theory of the Multiplet Effect

We saw in Section 10-9 how multiplet effects result from coupled spin states that become polarized. The type of multiplet effect (E/A or A/E) can be predicted in much the same way as the net effect [Eq. (10-2)], except that more variables must be considered. The multiplet effect (Γ_m) is given by the sign of the product

MULTIPLET EFFECT $\qquad\qquad \Gamma_m = (\mu)(\varepsilon)(a_i)(a_j)(J_{ij})(\sigma)$ $\qquad\qquad$ **(10-3)**

where μ and ε have the same meaning as before, a_i and a_j are the hyperfine couplings of the electron to nuclei i and j, and J_{ij} is the coupling constant they share. The sign of σ is positive if nuclei i and j are in the same radical, negative if they are in separate radicals.

- A positive Γ_m implies an E/A multiplet effect; a negative Γ_m corresponds to an A/E effect (Figure 10-5).

Note that Γ_m does *not* involve g values, so even if the two radicals in the pair are identical, they can still generate a multiplet effect.

EXAMPLE 10-11 When symmetrical peroxide **10-8** is thermally decomposed in a chlorinated solvent (abbreviated S-Cl), the major products beside CO_2 are compounds **10-9** and **10-10**.[3]

$$CH_3CH_2-\overset{\overset{\displaystyle O}{\|}}{C}-O-O-\overset{\overset{\displaystyle O}{\|}}{C}-CH_2CH_3 \xrightarrow{\text{S-Cl}} 2\,CO_2 + CH_3CH_2CH_2CH_3 + CH_3CH_2Cl$$

$\qquad\qquad$ **10-8** $\qquad\qquad\qquad\qquad\qquad\qquad\qquad$ **10-9** $\qquad\qquad$ **10-10**

Predict the type of ^1H multiplet effect expected for products **10-9** and **10-10**.

Solution: From our discussion in Example 10-10, we can assume that the initial radical pair (**10-11**) is "born" in the singlet state.

$$10\text{-}8 \longrightarrow \overline{CH_3CH_2\cdot \quad \cdot CH_2CH_3} + 2\,CO_2$$

$$\textbf{10-11}$$

recombination   S-Cl
escape

10-9   **10-10**

Furthermore, it is reasonable to expect that **10-9** arises from recombination within the cage, while **10-10** is an escape product. The ^{1}H spectrum of **10-10** consists of the familiar (Chapter 8) quartet and triplet. With regard to the precursor ethyl radicals, we know (Section 10-5) that a_α is negative and a_β is positive, while (from Section 9-2) $J_{\alpha\beta}$ is positive. And since both H_α and H_β occur in the same radical, σ is positive.

$$\begin{array}{c} H_\alpha \quad H_\beta \\ | \qquad | \\ \cdot C - C - H_\beta \\ | \qquad | \\ H_\alpha \quad H_\beta \end{array}$$

ethyl radical

10-11

Thus, for the CH_2 quartet of **10-10** we predict

$$\Gamma_m = (-)(-)(-)(+)(+)(+) = (-)$$

for an A/E multiplet effect. For the CH_3 triplet,

$$\Gamma_m = (-)(-)(+)(-)(+)(+) = (-)$$

and again an A/E multiplet effect is predicted. In perfect agreement with our expectations, both multiplets of **10-10** exhibit A/E multiplet effects.[3]

In the case of recombination product **10-9**, the only difference is that ε is now positive. Therefore, both multiplets are predicted to show E/A multiplet effects, and they do!

10-12. A Few Final Words about CIDNP

In many cases an nmr spectrum exhibits *both* net and multiplet CIDNP effects. In many of these cases the net E or A effect is superimposed on the multiplet effect of a given signal. For example, when singlet radical pair **10-12** undergoes recombination to **10-13**, the ^{1}H spectrum of the CH_2 group in **10-13** shows an E/A multiplet effect superimposed on an E net effect.[2]

10-12   **10-13**

The net effect can be rationalized by Eq. (10-2), $\Gamma_n = \mu\varepsilon\,\Delta ga$. Recalling that the value of g for $CH_3CH_2\cdot$ is less than that of $RCO_2\cdot$ (Table 10-1), we calculate for the CH_2 quartet

$$\Gamma_n = \mu\varepsilon\,\Delta ga = (-)(+)(-)(-) = (-)$$

for a net E effect. The E/A multiplet effect in the quartet can be predicted from Eq. (10-3).

$$\Gamma_m = (\mu)(\varepsilon)(a_i)(a_j)(J_{ij})(\sigma) = (-)(+)(-)(+)(+)(+) = (+)$$

for an E/A multiplet effect.

And finally, as hinted in Example 10-10b, CIDNP effects are not limited to ^{1}H spectra. Here is a case involving ^{13}C nmr.

EXAMPLE 10-12 The photolytic (light-induced) decomposition of compound **10-14** in CCl_4 is believed to involve singlet radical pair **10-15**, which undergoes cage recombination to product **10-16**.

$$RO_2C\text{—}CHN_2 \xrightarrow[CCl_4]{hv} \overline{RO_2C\text{—}\dot{C}H + N_2} \xrightarrow{-N_2} \overline{RO_2C\text{—}\dot{C}H + Cl\text{—}CCl_3}$$

10-14

$$\underset{\underset{Cl}{|}}{RO_2C\overset{1}{\text{—}}\overset{2}{CH}\overset{3}{\text{—}}CCl_3} \longleftarrow \underset{\underset{Cl}{|}}{\overline{RO_2C\text{—}CH\cdot \quad \cdot CCl_3}}$$

10-16   **10-15**

A net effect (E or A) was observed in each ^{13}C signal of product **10-16**. Predict the type of net effect for each carbon.

Solution: To calculate Γ_n for each *carbon* signal, we need to know (1) the sign of Δg (g values: $\cdot CCl_3$, 2.0091; $RO_2C\text{—}CHCl$, 2.003) and (2) the signs of the hyperfine couplings to the *carbons* (a_1 negative, a_2 positive, a_3 positive). With these data, and the fact that we're dealing with cage recombination ($\varepsilon = +$) of a singlet radical pair ($\mu = -$), we predict the following net effects:

Carbon	Γ_n	Net effect
1	$(-)(+)(-)(-) = (-)$	E
2	$(-)(+)(-)(+) = (+)$	A
3	$(-)(+)(+)(+) = (-)$	E

And, believe it or not, this is exactly what was observed![4]

The main thing to remember from this discussion is that CIDNP effects are seen in the nmr spectra of the *products* of reactions involving radical pairs. You *don't* observe the radicals themselves, as you do with esr spectroscopy. So, the next time your nmr spectrometer shows peaks that are either too big, negative, or strangely constructed, don't automatically assume your instrument is on the blink. Perhaps the spectrum is trying to tell you something!

REFERENCES

1. D. J. Pasto and C. R. Johnson, *Organic Structure Determination*. Prentice-Hall, New York, 1969.

2. H. R. Ward, *Acc. Chem. Res.* **5**, 18 (1972).

3. S. H. Pine, *J. Chem. Educ.* **49**, 664 (1972).

4. H. Iwamura, Y. Imahashi, and K. Kushida, *J. Am. Chem. Soc.* **96**, 921 (1974).

SUMMARY

1. Molecules or molecular fragments that possess one (or more) *unpaired* electrons are called free radicals. Free radicals are usually formed through cleavage of a chemical bond; recombination describes the process of two free radicals combining to re-form the bond.

2. An unpaired electron, like a proton, can adopt either of two spin orientations when immersed in a magnetic field. An electron in either orientation will precess at a frequency given by

$$\nu_{prec} = (constant)gB_0 \tag{10-1}$$

   The Landé factor g can be compared to the magnetogyric ratio used in nmr spectroscopy. The exact value of g depends on the structure of the free radical possessing the upaired electron.

3. Unpaired electrons precess about 650 times faster than protons do at the same magnetic field

strength. The energy gap between the two electronic spin states is also 650 times larger than that for protons, which translates into a difference in populations of the two spin states greater than in nmr.

4. An electron spin resonance (esr) or electron paramagnetic resonance (epr) signal is generated when an unpaired electron in the lower-energy spin state absorbs a photon at v_{prec} and is elevated to the higher-energy spin state. Esr signals are usually displayed in the dispersion mode.

5. Esr signals, like nmr signals, can be split into multiplets by coupling between the unpaired electron and neighboring magnetic nuclei. Such splitting is governed by the same type of first-order coupling rules that govern internuclear spin coupling, except that the hyperfine coupling constant (*a* value) is expressed in gauss rather than in hertz.

6. A radical pair (two interacting radicals) can exist in either the singlet state (electronic spins paired) or triplet state (with electronic spins parallel).

7. Nmr spectra of compounds undergoing reactions involving radical pairs often exhibit anomalous signals as the result of chemically induced dynamic nuclear polarization (CIDNP). While esr spectral signals are due to radicals themselves, CIDNP signals are due to the *products* arising from radical-pair reactions. These products can be formed with nuclei in polarized (nonequilibrium) spin state distributions.

8. CIDNP signals are of two basic types:

   (a) Those exhibiting a NET effect: highly intense positive (net absorption, or *A*) signals, or intense negative (upside-down, net emission, or *E*) signals.

   (b) Those exhibiting a MULTIPLET effect: both positive and negative signals in the same multiplet. Multiplet effects are designated *A/E* or *E/A*.

9. The direction (positive or negative) of the net effect (Γ_n) is determined by the sign of the product of four spectral variables:

$$\Gamma_n = (\mu)(\varepsilon)(\Delta g)(a) \tag{10-2}$$

Positive Γ_n, *A* net effect; negative Γ_n, *E* net effect.

10. The type of multiplet effect (*A/E* or *E/A*) is determined by the sign of the product of six spectral variables:

$$\Gamma_m = (\mu)(\varepsilon)(a_i)(a_j)(J_{ij})(\sigma) \tag{10-3}$$

Positive Γ_m, *E/A* multiplet effect; negative Γ_m, *A/E* multiplet effect.

11. The principal value of CIDNP spectra is that they supply information about the spin state of the radical pair at its birth, μ in Eqs. (10-2) and (10-3).

11 NUCLEAR MAGNETIC DOUBLE RESONANCE, PULSE SEQUENCES, AND 2D NMR

THIS CHAPTER IS ABOUT

- ☑ **An Introduction to Double Resonance Techniques**
- ☑ **Heteronuclear Spin Decoupling**
- ☑ **Heteronuclear NOE**
- ☑ **Off-Resonance Decoupling**
- ☑ **Homonuclear Spin Decoupling**
- ☑ **Homonuclear NOE**
- ☑ **Other Homonuclear Double Resonance Techniques**
- ☑ **Modern Pulse Sequence Techniques**
- ☑ **2D NMR**

11-1. An Introduction to Double Resonance Techniques

In Section 3-1 we examined the basic components of an nmr spectrometer. One of these essential components is the transmitter coil, through which radio-frequency alternating current flows (either continuously or in pulses) in order to irradiate the sample with an rf electromagnetic field. We are now going to make a significant addition to our spectrometer: a second transmitter coil which, like the first one, has its axis in the x,y plane (Figure 3-1). The purpose of the second transmitter coil is to enable us to bring one set of nuclei into resonance while we observe the effect of those nuclei on the signals of another set of nuclei. Because these two sets of nuclei undergo resonance simultaneously (though at different frequencies, by virtue of their different γ values), we refer to such experiments as **double resonance techniques**. To distinguish between the two rf fields, we will refer to the original one as the **observing** (or sensing) **field** because it is the one that generates the nmr signal of interest. The other (new) rf field is called the **irradiating field**, even though *both* rf fields actually irradiate the sample. Each field has its own frequency, and we'll refer to these as the **observing frequency** (v_1) and the **irradiating frequency** (v_2).

When the two sets of nuclei are of different types (i.e., different isotopes of an element, or different elements) we speak of *heteronuclear* double resonance. In this case, v_1 and v_2 are generated independently by two separately tunable rf oscillators. When we *observe* the signal from nuclei of isotope A while *irradiating* the nuclei of isotope B, we label the resulting spectrum with the shorthand designation A{B}. When the two sets of nuclei belong to the same isotope (e.g., both ^1H), we describe the technique as *homonuclear* double resonance.

EXAMPLE 11-1 Suppose we were to examine the ^1H spectrum of CH_2F_2 (Section 8-6A) at a field strength of 2.35 T, while simultaneously irradiating the fluorines. **(a)** Does this experiment involve homo- or heteronuclear double resonance? **(b)** What are the values of v_1, v_2, and the difference (Δv) between them?

Solution:

(a) Heteronuclear, because ^1H and ^{19}F are different elements. We would designate the resulting spectrum with the label $^1H\{^{19}F\}$.

(b) The ^{1}H frequency (v_1) would be 100 MHz (Example 2-6), while the ^{19}F frequency (v_2) is 94 MHz, as calculated from $v_{\text{precession}} = \gamma B_0/2\pi$ [Eq. (2-6)] and the γ value for ^{19}F in Table 2-1. Therefore, $\Delta v = 100$ MHz $- 94$ MHz $= 6$ MHz $= 6{,}000{,}000$ Hz.

EXAMPLE 11-2 Suppose we were to reexamine the 60-MHz ^{1}H spectrum of diethyl ether (Section 8-2 and Figure 8-1). But this time we *observe* the methyl triplet while we *irradiate* the methylene quartet. **(a)** Does this experiment involve homo- or heteronuclear double resonance? **(b)** What is the value of Δv in this case?

Solution:

    (a) Homonuclear, because both sets of nuclei are ^{1}H.
    (b) From Figure 8-1 (or Example 8-2), we can see that the triplet is separated from the quartet by 135 Hz.

These two examples show that the *difference* between v_1 and v_2 is very much *smaller* in the case of homonuclear double resonance, suggesting a simpler way to generate v_2. Instead of using a second rf oscillator, we can use the same one that provides v_1 but electronically modulate part of its output with an audio-frequency signal to give v_2. This technique makes it easier to control the exact value of Δv and allows us to focus v_1 and v_2 exactly where we want them.

The family of double resonance techniques has many members and in this chapter we'll examine some of the most common. We'll discover that the different effects of double resonance depend primarily on the exact location, timing, and power of the irradiating field. Furthermore, by careful tuning of the observing transmitter coil, one can send *both* v_1 and v_2 through it, obviating the need for the second transmitter coil. It is even possible to add one (or more!) additional irradiating fields, but these higher-order multiple resonance techniques are beyond the scope of this book.

You might wonder what would happen if v_1 and v_2 accidentally took on the same value. Suppose v_2 were set on the quartet of diethyl ether (Figure 8-1) while v_1 was scanned through the entire spectrum under continuous-wave conditions. As v_1 scanned through v_2, a beat pattern (Figure 11-1) would be generated at the position of the quartet by interference between the two frequencies. Compare this beat pattern with the ringing effect caused when the frequency is scanned rapidly through an nmr signal under continuous-wave conditions (Section 3-4A and Figure 3-14). Under PFT conditions, the Fourier transformation usually replaces such a beat pattern with a negative peak.

FIGURE 11-1 A beat pattern resulting from interference between two frequencies.

11-2. Heteronuclear Spin Decoupling

When we investigated heteronuclear ^{13}C$-$H spin coupling (Section 8-6A), we learned that all the ^{13}C spectra presented in Chapters 5 and 7 had been *spin decoupled*, because in the absence of such decoupling the spectra would have been extremely complex. But how do we uncouple the ^{1}H nuclei when generating a ^{13}C spectrum? By using a ^{13}C$\{^1$H$\}$ technique! While *observing* spectral data at the precessional frequency of ^{13}C (v_1), we simultaneously subject the sample to a powerful *irradiating* field adjusted to the average precessional frequency of the ^{1}H nuclei (v_2). Actually, the irradiating field is a **broadband** composed of all the frequencies in the ^{1}H region, much like the pulse used in the Fourier transform technique (Section 3-3 and Figure 3-8), and as such is often described as **white noise**.

This irradiation of all the protons at their resonance frequencies causes them to precess in bundles, with a component of magnetization appearing in the x,y plane (Figure 2-7) and a reduced component in the z direction. If the irradiating field is strong enough, not only will the ^{1}H nuclei approach saturation, but virtually all the ^{1}H magnetization will be tipped into the x,y plane. Since the ^{1}H nuclei are no longer aligned with (or against) the applied field (which is along the z axis), they can no longer augment or diminish the magnetic field experienced by the carbons (Section 8-3). As a result, the coupling interaction disappears, and each ^{13}C multiplet collapses to a singlet!

Example 8-13 described what the ^{13}C spectrum of 2-chlorobutane would have looked like in the *absence* of $\{^{1}H\}$ decoupling. Figure 7-1 showed the proton-noise-decoupled ^{13}C spectrum of the compound, with one sharp line for each carbon. Clearly, the main virtue of heteronuclear spin decoupling is its simplifying effect on the appearance of the spectrum, making the task of measuring chemical shifts and assigning signals much easier. But there is an additional benefit as well, an improvement in the signal-to-noise ratio (Section 3-2C). One reason for this is that the several lines of each multiplet are now collected into a single line with the same total intensity. But there is another, more subtle, intensity-increasing factor called the **nuclear Overhauser effect** (NOE), first mentioned in Section 5-4A.

11-3. Heteronuclear NOE

In Section 10-8 we saw how polarization (departure from the equilibrium distribution) of spin state populations can have a significant effect on nmr signal intensity. It turns out that in a $^{13}C\{^{1}H\}$ experiment, irradiation of the ^{1}H nuclei affects the nuclear spin state populations in such a way as to increase the intensity of the ^{13}C signals. Let's see how this can happen.

First of all, the NOE interaction between two magnetic dipoles takes place predominantly through *space* (i.e., directly), rather than through *bonds* (as is the case in normal spin coupling). An interaction

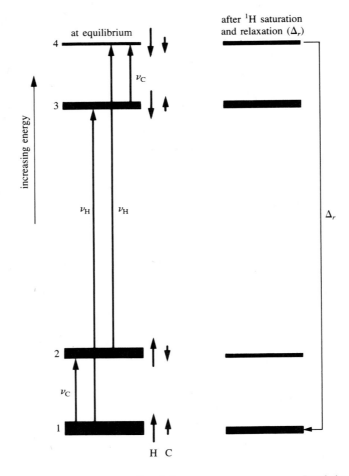

State	Population	Population Difference	Population	Population Difference
4	$-\Delta_H - \Delta_C$ ⎱		$-\Delta_C - \Delta_r$ ⎱	
3	$-\Delta_H + \Delta_C$ ⎰	$2\Delta_C$	$+\Delta_C$ ⎰	$2\Delta_C + \Delta_r$
2	$+\Delta_H - \Delta_C$ ⎱		$-\Delta_C$ ⎱	
1	$+\Delta_H + \Delta_C$ ⎰	$2\Delta_C$	$+\Delta_C + \Delta_r$ ⎰	$2\Delta_C + \Delta_r$

FIGURE 11-2 Spin state populations resulting in the nuclear Overhauser effect.

through space is very sensitive to the distance (r) between the interacting nuclei, being inversely proportional to r^3. Thus, for two nuclei to exhibit the NOE, they must be close together in the molecular structure. They need not be spin coupled, but the NOE is not precluded if they are.

Consider a molecule that possesses a ^{13}C atom and a nearby (though not necessarily spin coupled) ^1H atom. The combinations of spin states for this two-spin system shown in Figure 11-2 are quite similar to the spin states of two ^1H nuclei (Figure 9-1), except that ΔE for the ^1H nuclei is four times that for the ^{13}C nuclei (Example 2-5). As we did in Chapter 10, we'll depict relative spin state populations by the thickness of the spin state line. Recall from Section 2-3A that at equilibrium there will be a slight preference for nuclei to be in their lower-energy spin state (parallel to the applied field). We'll call this preference $2\Delta_C$ for the carbon and $2\Delta_H$ for the hydrogen. Therefore, the population of state 1 is augmented by $\Delta_H + \Delta_C$, state 2 by $\Delta_H - \Delta_C$, etc.

The ^{13}C signal results from transitions between states 1 and 2 and between states 3 and 4. Its intensity is a function of the *difference* in populations between states 1 and 2 and between states 3 and 4. At equilibrium, this population difference is $2\Delta_C$ in both cases. Now, if the ^1H states were to become saturated in a double resonance experiment, Δ_H would become zero. This would mean that states 1 and 3 would become equally populated, as would states 2 and 4. At first glance it would seem that the population *difference* between states 1 and 2 and between states 3 and 4 would still be $2\Delta_C$, but another factor intervenes. Notice that the populations of states 1 and 4 are polarized. Before saturation, the difference between them was $2\Delta_H + 2\Delta_C$, but at saturation it is only $2\Delta_C$. In an attempt to restore equilibrium, a number (Δ_r) of nuclei in state 4 relax to state 1, provided that we allow enough time for this relaxation to evolve. Thus, the population of state 1 will increase (by Δ_r), while that of state 4 will decrease (by Δ_r), and the intensity of both ^{13}C lines will increase in proportion.[1] The NOE is an example of **cross polarization**, because polarization of one set of nuclear spin states (here the saturation of the ^1H nuclei) results in the polarization of another set (here the ^{13}C nuclear spin states).

The maximum Overhauser signal enhancement (η) is given by

NOE ENHANCEMENT
$$\eta = \frac{1}{2}\left(\frac{\gamma_{irr}}{\gamma_{obs}}\right)$$
(11-1)

The γ values can be found in Table 2-1. Notice that a η value of 1.0 indicates a 100% enhancement, to give a signal twice what it was without the NOE.

EXAMPLE 11-3 Calculate the maximum NOE signal enhancement of a $^{13}C\{^1H\}$ signal.

Solution: Using Eq. (11-1) and the γ values for ^{13}C and ^1H in Table 2-1, we find

$$\eta = \frac{1}{2}\left(\frac{\gamma_{irr}}{\gamma_{obs}}\right) = \frac{1}{2}\left(\frac{267.5 \times 10^6}{67.26 \times 10^6}\right) = 1.988$$

Thus, a signal from a ^{13}C nucleus with a neighboring ^1H nucleus can be enhanced by nearly 200%, giving a signal three times as intense as one lacking the NOE, i.e., a ^{13}C nucleus with no nearby hydrogens.

The total NOE enhancement for a given carbon signal increases with the number of nearby hydrogens (though not in direct proportion). This is why we saw in Sections 5-4A and 7-2 that ^{13}C signal intensities usually follow the order methyl ($-CH_3$) > methylene ($-CH_2-$) > methine ($-\overset{|}{\underset{|}{C}}H$) > quaternary carbon ($-\overset{|}{\underset{|}{C}}-$).

To summarize, the NOE is an added bonus we get when generating a proton-decoupled ^{13}C spectrum. It helps overcome the low natural abundance and low sensitivity of the ^{13}C nucleus (Table 2-1). But remember, it works best when the *more* abundant nucleus (e.g., ^1H) is irradiated and the effect is observed on the *less* abundant nucleus (e.g., ^{13}C). It would be fruitless (as well as unnecessary) to carry out the reverse experiment, $^1H\{^{13}C\}$, and expect to see any enhancement of the hydrogen signal.

We'll discuss *homonuclear* NOE in Section 11-6, but there is one aside to mention at this point. What would happen if, instead of one *nucleus* polarizing the spin states of another nucleus, an *unpaired electron* polarized the spin states of a nearby nucleus? If you read Chapter 10, you know that this is related to what happens in the case of CIDNP!

EXAMPLE 11-4 Calculate the maximum signal enhancement due to the Overhauser effect between an unpaired electron and a nucleus.

Solution: From Section 10-4 we know that the ratio γ_e/γ_H is 655. Therefore,

$$\eta = \frac{1}{2}\left(\frac{\gamma_{irr}}{\gamma_{obs}}\right) = \frac{1}{2}\left(\frac{\gamma_e}{\gamma_H}\right) = \frac{1}{2}(655) = 328$$

This is in part why CIDNP-enhanced absorptions can be so intense.

11-4. Off-Resonance Decoupling

Every silver lining has a cloud! Although an ^{1}H-decoupled ^{13}C spectrum has the advantages of simplicity and improved signal-to-noise ratio, something has been lost. That "something" is the information that coupling provides about the structure of the sample (Chapters 8 and 9). For example, in a *coupled* ^{13}C spectrum (considering only one-bond C–H couplings), methyl carbons would appear as quartets, methylenes as triplets, etc., making them easier to assign. In fully decoupled spectra, all the signals appear as singlets. As we saw in Chapter 7, chemical shift calculations and NOE enhancements are not always definitive in allowing us to differentiate between carbon signals, especially when we study highly complex molecules. Wouldn't it be great if we could turn off all proton–carbon coupling *except* between hydrogens directly attached (one-bond separation) to carbons? In essence, we can do just that!

The technique is called **off-resonance decoupling**, and, as the term implies, we set v_2 not at the center of the ^{1}H resonance frequency, but about 1000–3000 Hz away. While this is still close enough to preserve most of the NOE enhancement of the ^{13}C signals, all coupling constants (J) are reduced in magnitude according to

REDUCED COUPLING CONSTANT $$J_r = \frac{2\pi J\,\Delta v}{\gamma_H B_2}$$ (11-2)

where J_r is the **reduced** (observed, or sometimes called *residual*) **coupling constant**, Δv is the difference between v_2 and v_{prec} of the coupling ^{1}H nucleus, and B_2 is the strength of the irradiating rf field (typically $\sim 2 \times 10^{-4}$ T). Notice that if $v_2 = v_{prec}$ (and B_2 is strong enough), all ^{13}C–H coupling constants are reduced to zero (complete decoupling).

EXAMPLE 11-5 Typical ^{13}C–H coupling constants (Sections 9-3 to 9-5) are ^1J, 125 Hz; ^2J, 5 Hz; ^3J, 5 Hz. Calculate the reduced coupling constants in an off-resonance decoupled spectrum where Δv averages 2000 Hz and $B_2 = 2 \times 10^{-4}$ T.

Solution: Apply Eq. (11-2):

$$^1J_r = \frac{2\pi J\,\Delta v}{\gamma_H B_2} = \frac{2\pi(125\ \text{Hz})(2000\ \text{Hz})}{(267.5 \times 10^6\ \text{rad T}^{-1}\ \text{Hz})(2 \times 10^{-4}\ \text{T})} = 29\ \text{Hz}$$

In like manner, $^2J_r = {}^3J_r = 1$ Hz, which is negligibly small. So, only one-bond C–H couplings will be observed.

EXAMPLE 11-6 Predict the appearance of the off-resonance-decoupled ^{13}C spectrum of 2-chlorobutane (Examples 7-3 and 8-13).

$$\underset{a}{\underbrace{H_3C}}\!-\!\overset{\displaystyle \overset{Cl}{|}}{\underset{b}{\underbrace{CH}}}\!-\!\underset{c}{\underbrace{CH_2}}\!-\!\underset{d}{\underbrace{CH_3}}$$

11-1

Solution: Deleting all but one-bond ^{13}C–H couplings leaves C_a–H_a, C_b–H_b, C_c–H_c, and C_d–H_d as the only couplings. Using the chemical shift data in Example 7-3, we therefore predict a quartet at δ 25.0 from methyl carbon C_a, a doublet at δ 60.1 from methine carbon C_b, a triplet at δ 33.6 from methylene carbon C_c, and a quartet at δ 11.1 from methyl carbon C_d.

As useful as off-resonance decoupling once was, there is an even better way to get the same type of information, as we'll see in Section 11-8.

11-5. Homonuclear Spin Decoupling

It would be wonderful if we could use a double resonance technique to generate an ^1H spectrum free of all homonuclear coupling, so that each set of hydrogens gave a singlet. But, of course, if we were to subject all the hydrogens to broadband irradiation (v_2) and try to observe their signals simultaneously, v_1 would equal v_2 and a horrendous mush of overlapping beat patterns (Section 11-1) would result. Nonetheless, we *can* carry out *partial* decoupling by focusing v_2 on one ^1H signal and observing the effect on the rest of the ^1H spectrum. This technique, **homonuclear spin decoupling**, results in partial simplification of the spectrum and confirms which nuclei are coupled to which others.

Take a moment to review the ^1H spectrum of structure **11-1** (Example 8-11, reproduced here as Figure 11-3). What would be the effect of irradiating the complex multiplet at δ 1.67 ppm? Saturating the

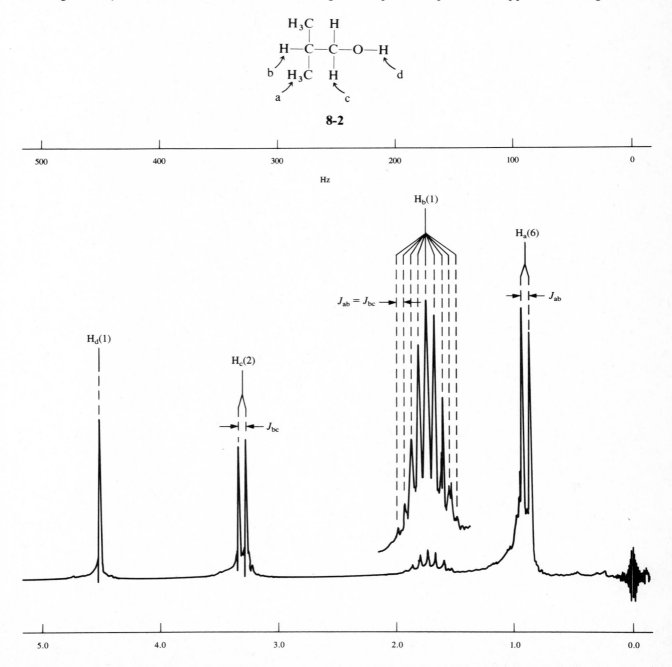

FIGURE 11-3 The 100-MHz ^1H spectrum of isobutyl alcohol, $(CH_3)_2CHCH_2OH$.

spin states of H_b by irradiation at that frequency would cause all spin coupling to that nucleus to disappear. As a result, the doublets centered at δ 0.89 and 3.27 ppm would both collapse to singlets, and this is exactly what is observed!

There are, however, certain operational problems with the technique. First, it can be quite tedious and time-consuming to locate v_2 at exactly the desired frequency, then adjust its power to saturate only *that* signal without disturbing any nearby signals. Second, to confirm *all* couplings in a molecule with n sets of mutually coupled nuclei, it would be necessary to generate at least $n - 1$ separate decoupled spectra. Fortunately, there are easier ways to confirm which sets of nuclei are coupled, and we'll discuss these in Section 11-9.

EXAMPLE 11-7 (a) How many decoupled spectra would have to be run to confirm all couplings among three sets of mutually coupled nuclei? (b) What would be the effect on the ^1H spectrum of $C_6H_4FNO_2$ (Figure 11-4) of irradiating the signal at δ 7.19 ppm? At 8.20 ppm? The fluorine signal? (See Example 8-19.)

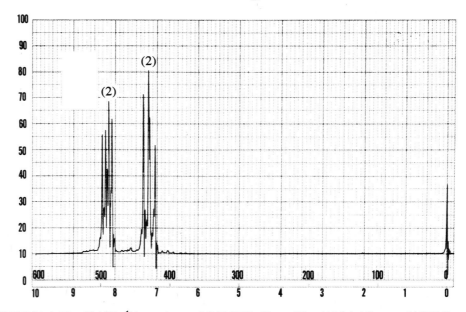

FIGURE 11-4 The 60-MHz ^1H spectrum of $C_6H_4FNO_2$ (From *The Aldrich Library of NMR Spectra*).

Solution:

(a) Since $n = 3$, we'll need to run at least $n - 1 = 2$ separate spectra.

(b) The signal at δ 7.19 ppm is due to H_a, and irradiating it would cause the *partial* collapse of the H_b signal (δ 8.20 ppm) to a *doublet*, because coupling between H_b and F ($^4J = 5$ Hz) will still be present. Irradiation of the H_b signal would cause the H_a signal to collapse to a *doublet*, again because coupling of H_a to F ($^3J = 8$ Hz) is unaffected. In practice, it would be somewhat difficult to perform these two experiments because the two multiplets are so close. Irradiation of the fluorine signal would constitute *heteronuclear* decoupling ($^1H\{^{19}F\}$), and would result in the collapse of both ^1H multiplets to doublets, preserving $^3J_{HH}$ (9 Hz).

11-6. Homonuclear NOE

Just as saturation of ^{1}H spin states led to enhancement of signals for nearby ^{13}C nuclei through NOE cross polarization (Section 11-3), *homonuclear* double resonance can also affect signal *intensities* as well as multiplicities. As before, the effects are significant only if the nuclei are relatively close in space, as the following examples demonstrate. Furthermore, the magnitude of the effect depends on whether or not the nuclei are spin coupled (through bonds). If they are *not* spin coupled, a signal *enhancement* will occur; if they *are* spin coupled, the signal will be modestly *diminished* by the NOE.

EXAMPLE 11-8 What is the maximum NOE signal enhancement expected in homonuclear NOE?

Solution: Because both γ values are equal, $\eta = \dfrac{1}{2}\left(\dfrac{\gamma_{\text{irr}}}{\gamma_{\text{obs}}}\right)$ [Eq. (11-1)] reduces to $\eta = 0.5$ (a 50% signal enhancement).

The ^{1}H spectrum of dimethylformamide (Figure 11-5) consists of *two* methyl singlets (at δ 2.81 and 2.98 ppm), and the remaining (so-called formyl) hydrogen appears at δ 7.90 ppm. There is no observable spin coupling between any of the hydrogens. (The reason the two methyl groups are *not* rendered equivalent by rotation around the C$-$N bond is discussed in Section 12-2B.) The question is this: Which signal belongs to which methyl group?

O
$\diagdown$ CH$_3$ (trans)
    C$-$N
H $\diagup$ $^{\diagdown}$CH$_3$ (cis)

dimethylformamide

11-2

To carry out this experiment, the intensity of the formyl hydrogen signal can be observed (v_1) while first one, then the other, methyl signal is irradiated (v_2). When the signal at δ 2.81 ppm is irradiated, the formyl signal is essentially unaffected. But irradiation of the signal at δ 2.98 ppm leads to an 18% enhancement of the formyl signal.[2] This result indicates that the signal at δ 2.98 ppm belongs to the methyl group closer to the formyl hydrogen, the cis methyl group.

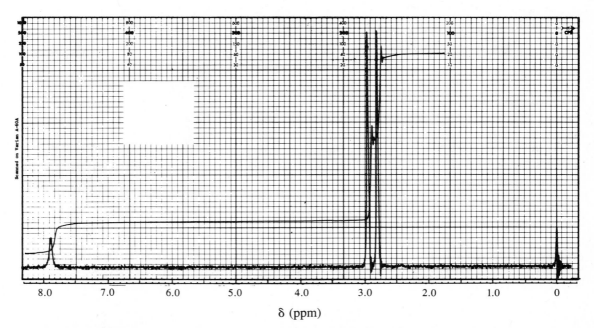

FIGURE 11-5 The 60-MHz ^{1}H spectrum of dimethylformamide. ©Sadtler Research Laboratories, Division of Bio-Rad Laboratories, Inc., 1967.

EXAMPLE 11-9 The ^{1}H spectrum of the compound with structure **11-3** consists of two nine-hydrogen singlets (δ 0.96 and 1.20 ppm) and two one-hydrogen singlets (O—H at δ 2.25 ppm and C—H at δ 3.94 ppm).[3]

11-3

While the intensity of the C—H signal was observed, each *tert*-butyl signal was (separately) irradiated. Irradiation of the signal at δ 1.20 ppm produced no significant change, but irradiation at δ 0.96 ppm gave an 11% *increase* in the intensity of the signal at δ 3.94 ppm. (**a**) Which *tert*-butyl group belongs to which signal? (**b**) Why not irradiate the C—H signal and examine the effect on the intensity of the *tert*-butyl signals?

Solution:

(**a**) These results indicate that the signal at δ 0.96 ppm belongs to the *tert*-butyl group *closer* to the C—H hydrogen, labeled 2 in structure **11-3**.

(**b**) Remember that the NOE is cumulative with the number of irradiated nuclei (Section 11-3). Therefore, it makes much more sense to monitor the effect that *nine* irradiated nuclei have on *one*, rather than the effect *one* irradiated nucleus has on *nine*.

11-7. Other Homonuclear Double Resonance Techniques

Most other homonuclear double resonance techniques, with names such as *spin tickling* and *selective decoupling*, provide information about the arrangement of spin states and the relationships between spin transitions. A comprehensive discussion of all these techniques is beyond the scope of this book, but we'll examine one of them in some detail.

In Section 9-1 we found that coupling constants have *signs* related to the ordering of spin states and spectral lines, a fact that usually has no effect on the appearance of a typical nmr spectrum. Whether these signs are the same or opposite can be determined by the **selective decoupling** technique. In this method only *part* of a given multiplet is irradiated, and decoupling is observed only in the "connected" *part* of each coupled multiplet. Let's see how this works.

Consider a three-spin AMX system, characterized by three chemical shifts (δ_A, δ_B, and δ_C) and three coupling constants (J_{AM}, J_{AX}, and J_{MX}).[4] The first-order (Section 8-5) spectrum of this system consists of three doublets of doublets, as shown in Figure 11-6. Each line of the A multiplet corresponds to a specific spin orientation of nuclei M and X; each line of the M multiplet corresponds to a specific spin orientation of nuclei A and X, etc. The specific spin orientations that correspond to a certain spectral line are determined by the signs of the coupling constants.

Assume for the moment that all three coupling constants in our AMX system are positive, and focus your attention on the A and M multiplets. Lines A_1 and A_2 correspond to the "up" (parallel to the applied field) orientation of nucleus X, while lines A_3 and A_4 correspond to the "down" orientation of X. Similarly, lines M_1 and M_2 correspond to the "up" orientation of nucleus X, while lines M_3 and M_4 correspond to the "down" orientation. We describe this situation by saying that the A_1–A_2 doublet is *connected* (via the similar spin state of nucleus X) to the M_1–M_2 doublet, as is the A_3–A_4 doublet to the M_3–M_4 doublet. Now, if we were to carry out a simultaneous irradiation of lines A_1 and A_2 only (by focusing v_2 between them and carefully adjusting its power), only the connected multiplet (lines M_1 and M_2) would collapse to a singlet. Lines M_3 and M_4 would remain intact.

Suppose instead that J_{MX} is *negative*, but the other two coupling constants remain positive. In this case, lines M_1 and M_2 correspond to the "down" orientation of X, while lines M_3 and M_4 correspond to the "up" orientation. Now doublet A_1–A_2 is connected to the M_3–M_4 doublet, and the A_3–A_4 doublet is connected to the M_1–M_2 doublet. This time, irradiation of lines A_1 and A_2 results in the collapse of lines

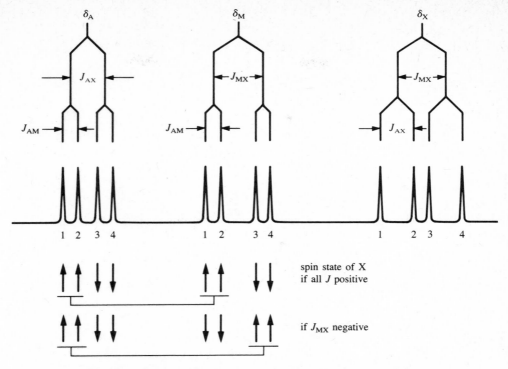

FIGURE 11-6 Connected multiplets in a three-spin AMX system.

M_3 and M_4. Thus, depending on which part of the M multiplet collapses when the A_1–A_2 doublet is irradiated, we can determine which pairs of lines are connected, and from this we can tell whether J_{MX} and J_{AX} have the same or opposite signs!

Through a similar series of experiments, we can determine the relative sign of the third coupling constant. But notice,

- Selective decoupling does not establish the *absolute* sign (plus or minus) of the coupling constants, only the relationship between their signs (same or opposite).

Usually, however, we can make an educated guess about the absolute sign of at least one of the coupling constants (Chapter 9) and use that to determine the absolute signs of the remaining ones.

EXAMPLE 11-10 Suppose J_{MX} and J_{AX} have opposite signs, while J_{AM} and J_{AX} have the same sign. **(a)** What is the sign relationship between J_{AM} and J_{MX}? **(b)** If J_{MX} is a one-bond C–H coupling constant, predict the absolute sign of all three J values.

Solution:

(a) Opposite signs.
(b) Since one-bond C–H coupling constants are positive (Section 9-3), J_{MX} is positive, while J_{AM} and J_{AX} are negative.

11-8. Modern Pulse Sequence Techniques

Before you get very far into these final two sections, you should review the concepts of pulsed Fourier transform nmr presented in Section 3-3. There we discussed the series of computer-controlled events that generate a PFT nmr spectrum: the pulse/data acquisition/delay sequence (hereafter simply called the pulse sequence), which is repeated some number of times, followed by the Fourier transformation of the acquired FID data from the time domain into the frequency domain. Recall that the duration and power of the initiating pulse determines the tip angle α of the net nuclear magnetization (Sections 2-3B and 3-3).

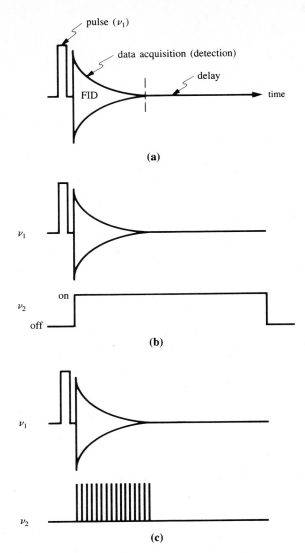

FIGURE 11-7 Some simple pulse sequences: **(a)** the normal detection pulse sequence; **(b)** detection with heteronuclear decoupling; **(c)** detection with homonuclear decoupling.

The usual pulse sequence shown in Figure 3-13 can be abbreviated as shown in Figure 11-7a. If we include a second frequency (v_2) for heteronuclear broadband spin decoupling, our representation would resemble Figure 11-7b. During homonuclear decoupling, the irradiating field can be turned off when each FID data point is collected, resulting in a spiked output of v_2 (Figure 11-7c).

As PFT instruments have become more widely available, scientists have discovered that altering the pulse sequence can lead to a large variety of interesting and useful effects. We'll examine just a few of these variations; for more detailed discussions see references 5 and 6 at the end of this chapter.

A. Gated decoupling

One relatively simple pulse sequence alteration, called **gated decoupling**, is depicted in Figure 11-8a. Because the NOE requires a certain amount of time for the critical relaxation to evolve (Section 11-3), its magnitude depends on the length of time v_2 is turned on. On the other hand, the decoupling effect of v_2 requires only that it be on during FID acquisition. Therefore, if we turn v_2 *on* during acquisition, but *off* during the subsequent delay time, decoupling would be preserved, but much of the NOE enhancement would be lost. By comparing the resulting spectrum with the usual one (produced by the pulse sequence in Figure 11-7b), the magnitude of NOE enhancement (η) for each signal can be determined, and this comparison provides valuable information about the arrangement of nuclei in a structure.

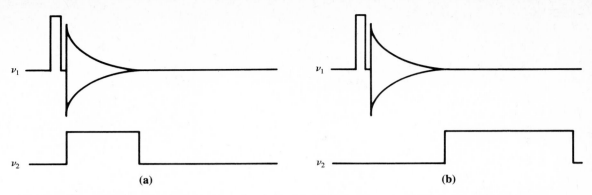

FIGURE 11-8 (a) The pulse sequence for gated decoupling. (b) The pulse sequence for inverse gated decoupling.

EXAMPLE 11-11 Inverse gated decoupling (Figure 11-8b) involves turning v_2 *off* during data acquisition, but *on* during the subsequent delay. What would be the effect of this pulse sequence?

Solution: There will be no decoupling (i.e., coupling will be preserved), but most of the NOE enhancement will be retained. The net result is a *coupled* spectrum with improved signal-to-noise ratio.

B. APT and SEFT pulse sequences

The majority of pulse sequence variations focus attention on the events that precede **detection** (data acquisition). The interval preceding detection is divided into a **preparation time**, the initiating pulse together with any prior relaxation delays, and an **evolution time**, during which additional timed pulses of v_1 and/or v_2 can occur. These relationships are depicted in Figure 11-9. Note that the evolution and detection times are given the symbols t_1 and t_2, not to be confused with the relaxation times T_1 and T_2 (Section 2-3B).

One very useful pulse sequence, known as the **spin echo technique**, is shown in Figure 11-10. The preparation period ends with a 90° pulse (Section 3-3), that tips the net nuclear magnetization **M** into the x,y plane. The evolution time is divided into two periods of equal length (τ) separated by a 180° pulse (which is twice as long as a 90° pulse).

note: Because the actual duration of a pulse ($\sim 10\ \mu s$) is negligible compared to τ (1–10 ms), $t_1 = 2\tau$.

During the first τ period each set of nuclei precesses at its own characteristic frequency, so their individual magnetic vectors (e.g., $\mathbf{M}_1$ and $\mathbf{M}_2$ in Figure 11-10b) begin to defocus (diverge). The effect of the 180° pulse is to reflect $\mathbf{M}_1$ and $\mathbf{M}_2$ through the y,z plane, so that the second τ period causes them to be refocused. Now, let's put this pulse sequence to good use.

Suppose we are examining the broadband ^{1}H-decoupled ^{13}C spectrum of a molecule that possesses a methyl (—CH_3) group, a methylene (—CH_2—) group, a methine ($-\overset{\displaystyle |}{\underset{\displaystyle |}{C}}$—H)

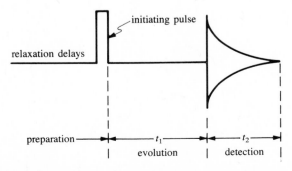

FIGURE 11-9 The evolution period in a pulse sequence.

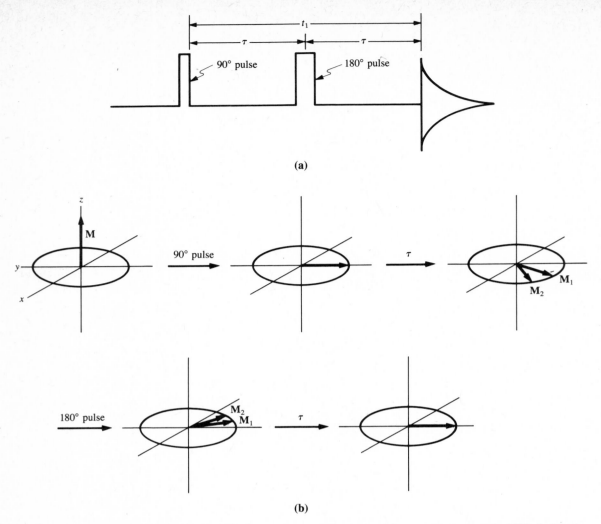

FIGURE 11-10 (a) The spin echo pulse sequence. (b) The evolution of nuclear magnetization (**M**) during the spin echo sequence.

group, and a quaternary carbon (—Ċ—). Furthermore, we'll assume that each of these

carbons precesses at its own characteristic frequency, and that all $^1J_{CH}$ values are comparable, ~ 125 Hz. In the *absence* of decoupling, the magnetization vectors (e.g., $\mathbf{M}_1$ and $\mathbf{M}_2$) for all but the quaternary carbon will be split by coupling to the attached hydrogen(s): the methine vector into two vectors (a doublet), the methylene vector into three (a triplet), and the methyl into four (a quartet). It turns out that if we gate (interrupt) the decoupling during *one* of the τ periods, the resulting "*J*-modulated" signals exhibit phases (and hence intensities) that are determined by their spin multiplicity and the relationship of τ to $1/J$. However, each signal still appears as a singlet because ν_2 is *on* during the detection phase.

There are two closely related forms of this experiment, APT (**attached proton test**,[5] Figure 11-11a) and SEFT (**spin echo Fourier transform**,[6] Figure 11-11b) that differ only in when ν_2 is on or off. In both cases, the intensity of each spectral signal varies from $+1$ to -1, depending on its multiplicity and the length of the evolution delay τ (in units of $1/J$), as shown in Figure 11-12. If we were to set $\tau = 0$, each signal would have intensity $+1$, and we would simply be collecting the normal $^{13}C\{^1H\}$ spectrum. By contrast, if we set $\tau = 4$ ms (i.e., $1/2J$), only the quaternary carbon signal would appear. All the other signals would have zero intensity because of destructive interference between the *J*-modulated vectors. If we set $\tau = 8$ ms ($= 1/J$), the quaternary and methylene carbons generate positive ($+1$) signals, while the methyl and methine carbons generate negative signals (-1). Thus, noting which signals go from positive to zero to negative as τ is varied enables us to assign

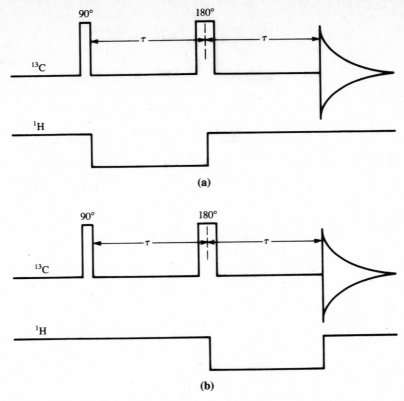

FIGURE 11-11 (a) The APT pulse sequence. (b) The SEFT pulse sequence.

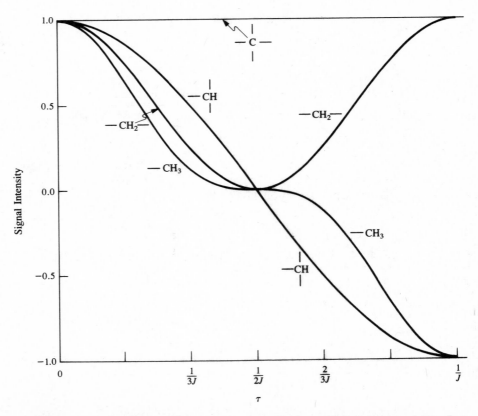

FIGURE 11-12 APT signal intensity as a function of spin multiplicity and evolution time.

unambiguously which signals are due to quaternary carbons, which are due to methylene carbons, and which are due to either methyl or methine carbons. Distinguishing the methyl and methine carbons is a little more difficult. If we set $\tau = 5.34$ ms ($= 2/3J$), however, both types of carbons will give negative signals, but the methine signal should be about four times as intense as the methyl signal. In principle it should be possible to assign all four multiplicities by the signals' relative intensities in one experiment where $\tau = 2/3J$. But in practice, interpreting small differences in relative intensities is more difficult than comparing large positive signals with large negative ones.

EXAMPLE 11-12 By filling in the following table, predict the relative intensities of each ^{13}C signal of 2-chlorobutane (Example 11-6) for each value of τ. You may assume that each $^1J_{CH}$ value is 125 Hz.

$$\begin{array}{c} \text{Cl} \\ | \\ \text{H}_3\text{C}-\text{CH}-\text{CH}_2-\text{CH}_3 \\ \uparrow \quad \uparrow \quad \uparrow \quad \uparrow \\ 25.0 \quad 60.1 \quad 33.6 \quad 11.1 \end{array}$$

τ (ms) \ δ_C (ppm)	11.1	25.0	33.6	60.1
0	+1	+1	+1	+1
4				
8				
5.34			—	

Solution:

τ \ δ	11.1	25.0	33.6	60.1	Remarks
0	+1	+1	+1	+1	All carbons
4 (1/2J)	0	0	0	0	None is quaternary
8 (1/J)	−1	−1	+1	−1	Only methylene carbon is +
5.34 (2/3J)	−0.1	−0.1	—	−0.4	First two are methyl carbons, fourth is a methine carbon

---

Although the APT and SEFT methods do require that three or four separate spectra be compared, the results (especially in the case of complicated molecules) are much more clear-cut than those of off-resonance decoupling (Section 11-4).

Some of the other types of pulse sequences that provide different types of structural information are listed in Table 11-1. They are referred to collectively as *one-dimensional* pulse

TABLE 11-1 Some Typical One-Dimensional Pulse Sequences

Acronym	Types of information provided
APT, SEFT	Multiplicity of ^{13}C Signal
SPI[a]	Signal assignments, relative signs of J values
INEPT[b]	Sensitivity enhancement; improved S/N
DEPT[c]	Sensitivity enhancement; improved S/N
INADEQUATE[d]	Determination of C–C coupling constants

[a] Selective Population Inversion
[b] Insensitive Nuclei Enhanced by Polarization Transfer
[c] Distortionless Enhancement by Polarization Transfer
[d] Incredible Natural Abundance Double Quantum Transfer

sequences for reasons that will become clear in the next section. For the scientists who develop these pulse sequences, the only thing more challenging than working out the details of the new technique is finding a name for it that provides a catchy acronym.

11-9. 2D NMR

A. General comments

The term **2D NMR**, which stands for two-dimensional nmr, is something of a misnomer. All the nmr spectra we've discussed so far in this book are two-dimensional in the sense that they are plots of signal intensity versus frequency (or its Fourier equivalent, signal intensity versus time). By contrast, 2DNMR refers to spectral data that are collected as a function of *two* time scales, t_1 (evolution) and t_2 (detection), as introduced in Section 11-8. So a 2D NMR spectrum is actually three-dimensional: signal intensity versus *two* frequencies, F_1 (the Fourier transform in the t_1 time domain) and F_2 (the Fourier transform in the t_2 time domain).

In APT or SEFT techniques (Section 11-8), we compare three or four spectra that differ in the length of the evolution period τ ($= t_1/2$) used in generating them. A 2D NMR spectrum is actually a large number of spectra (often several hundred) collected automatically, each with a slightly different (uniformly incremented) value of t_1. The resulting data are then Fourier-transformed in both the t_1 ($\rightarrow F_1$) and t_2 ($\rightarrow F_2$) domains. The resulting 2D NMR spectrum can be displayed in two ways: as a **stacked plot**, which portrays all three dimensions in a perspective view (Figure 11-13a), or as a **contour plot**, which represents a *horizontal* cross-section of the stacked plot at a selected value of signal intensity (Figure 11-13b).

As with one-dimensional techniques (Section 11-8), there is a large and growing family of 2DNMR pulse sequences,[5,6] each with its own catchy acronym. We'll look at only one of them in detail.

B. HETCOR 2D NMR

When attempting to unravel the structure of a complex unknown molecule from ^{13}C and ^1H spectral data, it is often quite helpful to know which ^1H signals are correlated (associated via spin coupling) with which ^{13}C signals, for this information can tell us which carbons are bonded to which hydrogens. We already know (Sections 5-5 and 7-7) that there is a rough parallel between the chemical shifts of carbons and the chemical shifts of the hydrogens attached to them. But in complex structures it can be difficult or impossible to apply this generalization because of the close proximity of so many signals. It's too bad that we can't persuade nmr spectrometers to display such correlations by printing color-coded signals. But a 2D NMR technique known as **heteroscalar correlation** (HETCOR) is the next best thing.

By applying the complicated pulse sequence shown in Figure 11-14, which involves cross polarization (Section 11-8) from ^1H to ^{13}C, we can generate a 2D spectrum in which the F_2 axis corresponds to ^{13}C chemical shifts (δ_C) and the F_1 axis corresponds to ^1H chemical shifts (δ_H). The keys to success with this technique are to set t_1 equal to $1/2J_{CH}$ and to use Δt as the evolution variable. The value picked for J_{CH} depends on the type of correlation desired.

Let's consider a specific example. The normal $^{13}C\{^1H\}$ spectrum of 2-chlorobutane (structure **11-4**) consists of signals at δ_C 11.1 (C_d), 25.0 (C_a), 33.6 (C_c), and 60.1 (C_b) ppm (Example 7-3). Its ^1H spectrum consists of signals at δ_H 1.01, 1.49, 1.72, and 3.88 ppm [we'll ignore the fact that the two H_c's are actually diastereotopic (Section 4-3)]. We can guess that these ^1H signals correspond to H_d, H_a, H_c, and H_b, respectively, but HETCOR 2D NMR allows us to confirm this fact unequivocally.

$$
\begin{array}{cccc}
H_a & Cl & H_c & H_d \\
| & | & | & | \\
H_a{-}C_a{-}C_b{-}C_c{-}C_d{-}H_d \\
| & | & | & | \\
H_a & H_b & H_c & H_d
\end{array}
$$

11-4

To select 125 Hz (a typical one-bond C–H coupling constant) as our value of J_{CH}, we set $t_1 = 4$ ms and generate a 2D NMR spectrum in which each signal corresponds to the intersection of a δ_C value with the δ_H value(s) of hydrogens to which it is coupled by 125 Hz. The resulting contour plot (Figure 11-15) confirms our tentative ^1H shift assignments, with signals correlating H_a (δ_H 1.01 ppm)

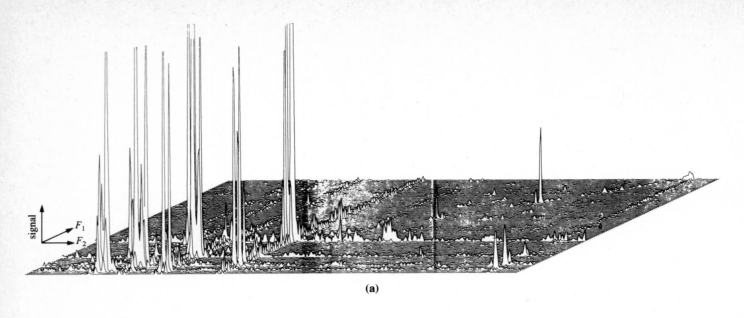

(a)

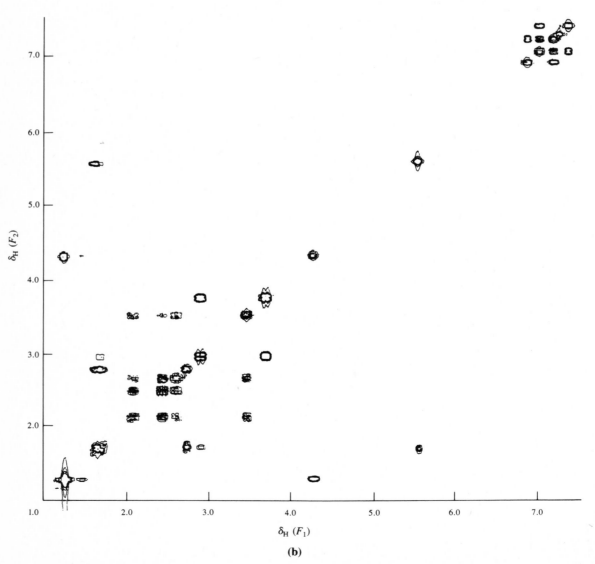

(b)

FIGURE 11-13 (**a**) A typical stacked 2D NMR plot. (**b**) A typical contour 2D NMR plot.

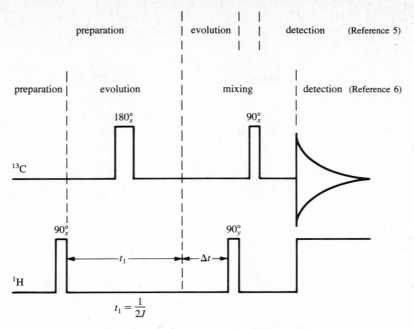

FIGURE 11-14 The HETCOR pulse sequence.

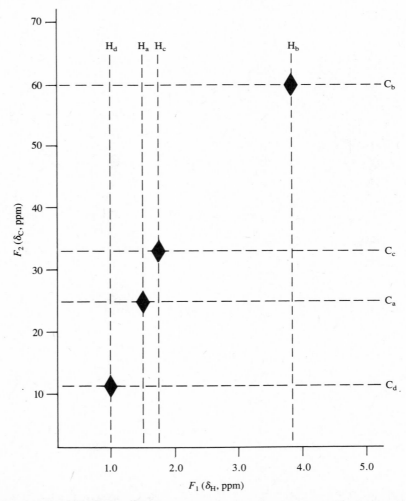

FIGURE 11-15 The idealized HETCOR 2D NMR spectrum of 2-chlorobutane, testing for one-bonded couplings.

and C_a (δ_C 11.1 ppm), H_b and C_b, etc. The diagonal relationship of the signals in this 2D NMR spectrum is a direct consequence of the parallel between carbon chemical shifts and attached hydrogen chemical shifts.

EXAMPLE 11-13 (a) What would be the result of repeating the HETCOR experiment on 2-chlorobutane with t_1 set to correspond to a J_{CH} value of 7 Hz (typical of two- and three-bond C–H coupling constants)? Use a diagram such as Figure 11-15 to explain your answer. **(b)** What is the required value of t_1?

Solution:

(a) See Figure 11-16. In this case, signals appear only at intersections between δ values that are coupled by a J value of 7 Hz, i.e., carbons and hydrogens that are separated by two or three bonds (Example 8-13).

(b) $t_1 = \dfrac{1}{2J} = \dfrac{1}{2(7 \text{ Hz})} = 71 \text{ ms}$

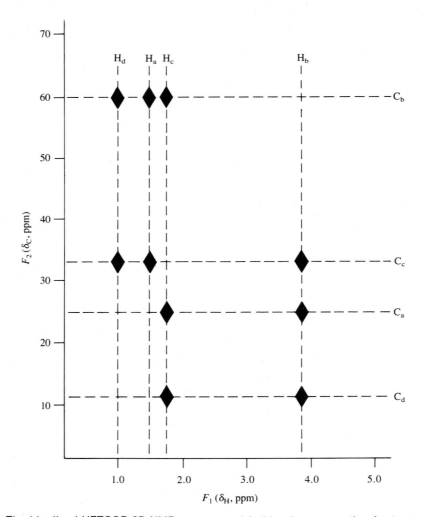

FIGURE 11-16 The idealized HETCOR 2D NMR spectrum of 2-chlorobutane, testing for two- and three-bond couplings.

A closely analogous *homonuclear* 2D NMR technique, known as HOMCOR or COSY **(correlated spectroscopy)**, maps out all homonuclear couplings in a *single* spectrum. Do you remember how many homonuclear decoupled spectra were required to get the same information

(Section 11-5)? In a typical COSY spectrum (of which Figures 11-13a and b are examples), both the F_1 and F_2 axes correspond to ^1H chemical shift scales. The normal ^1H spectrum appears along the diagonal, while all off-diagonal signals correspond to $^1H-^1H$ homonuclear couplings. (See Problem 9 in Self-Test II.) A 2D version of INADEQUATE (Section 11-8) is very useful in establishing which signals belong to carbons that are bonded directly to each other.

To summarize, the multitude of new pulse sequences greatly improves our ability to analyze complicated nmr spectra, to unravel ever more complex molecular structures, and to open doors to many new and exciting applications of nmr. We'll investigate some of these in Chapter 12.

REFERENCES

1. J. W. Cooper, *Spectroscopic Techniques for Organic Chemists.* Wiley, New York, 1980.
2. F. A. L. Anet and A. J. R. Bourn, *J. Am. Chem. Soc.* **87,** 5250 (1965).
3. R. C. Elder, L. R. Florian, E. R. Kennedy, and R. S. Macomber, *J. Org. Chem.* **38,** 4177 (1973).
4. E. Becker, *High Resolution NMR*, 2nd ed. Academic Press, New York, 1980.
5. J. N. Shoolery, *J. Nat. Prod.* **47,** 226 (1984).
6. R. Benn and H. Gunther, *Angew. Chem. Int. Ed. Engl.* **22,** 350 (1983).

SUMMARY

1. Double resonance nmr techniques involve simultaneous irradiation of the sample with electromagnetic radiation of two separate frequencies (v_1 and v_2). This allows us to determine the effect of irradiating (with v_2) one set of nuclei while observing (at v_1) the nmr signal of another set of nuclei.

2. *Homonuclear* double resonance indicates that both v_1 and v_2 are focused on nuclei of the same isotope, while *heteronuclear* double resonance indicates that the v_1 and v_2 are focused on different isotopes or elements.

3. Depending on the power and exact location of v_2, a variety of useful effects can be generated during double resonance. Some of the more common include

   (a) *Decoupling*: Irradiation of nuclei in set B, during observation of the signal from those in set A, can "erase" the effects of any A–B coupling. For example, irradiation of all ^1H nuclei during acquisition of a ^{13}C spectrum (denoted $^{13}C\{^1H\}$) results in a greatly simplified ^{13}C spectrum, free of all C–H couplings. Both homonuclear and heteronuclear decoupling are possible.

   (b) *Nuclear Overhauser effect (NOE)*: Irradiation of nuclei in set B can have an effect on the spin state populations of other nearby nuclei, causing a change in intensity of the signals of the nearby nuclei (usually an increase in intensity). Thus, in a $^{13}C\{^1H\}$ experiment, not only is the spectrum simplified by decoupling, but also the intensity of most signals is increased by the NOE.

   (c) *Off-resonance decoupling*: When v_2 is focused somewhat away from the resonance frequency of set B nuclei, while the spectrum of set A nuclei is observed, A–B coupling constants are reduced in magnitude but not eliminated completely. The reduced coupling constant (J_r) is given by

$$J_r = \frac{2\pi J\,\Delta v}{\gamma_B B_2} \tag{11-2}$$

   (d) *Selective decoupling*: By carefully adjusting the power and frequency of v_2, it is possible to decouple only parts of multiplets and thereby generate information that helps determine the relative signs of coupling constants.

4. Fourier transform nmr opened the door to development of techniques where the pulse–FID acquisition–delay sequence is altered and embellished in a variety of ways to give useful information about molecular structure. Such pulse sequence variations include

   (a) *Gated decoupling*: If v_2 is turned *on* during FID acquisition but *off* during the subsequent delay, much of the NOE enhancement is lost, though *de*coupling is preserved. This can give useful information about the relative location of nuclei within a molecule.

   (b) *Inverse gated decoupling*: If v_2 is *off* during acquisition, but *on* during the delay, the result is a spectrum where coupling is preserved and each signal enjoys the benefits of NOE enhancement.

(c) *Attached proton test (APT)* or *spin echo Fourier transform (SEFT)*: Applying a 180° refocusing pulse (v_1) midway through the evolution time makes possible a determination of how many hydrogens are attached to each carbon.

5. Certain pulse sequences can also provide FID data that can be Fourier transformed in two different dimensions (evolution and detection). The resulting spectra (or family of spectra) are termed two-dimensional (2D NMR). Examples of this technique include, among many others,

   (a) Heteroscalar correlated (HETCOR) spectra, which provide direct information regarding which nuclei are connected to which other nuclei.

   (b) Homoscalar (HOMCOR or COSY) spectra, which map out all homonuclear couplings.

12 SPECIAL TOPICS

THIS CHAPTER IS ABOUT

☑ **Chemical Shift Reagents**
☑ **The Study of Dynamic Processes by NMR**
☑ **CPMAS Solid-Phase NMR**
☑ **Surface Coil NMR and Magnetic Resonance Imaging**
☑ **Afterword**

12-1. Chemical Shift Reagents

In Section 9-9 we discovered that the appearance of a spin coupled nmr spectrum is determined by the ratio of Δv (the difference in chemical shifts between the coupling nuclei) to J (the coupling constant they share). For the spectrum to exhibit first-order (Pascal's triangle, Section 8-5) multiplet intensities, the value of $\Delta v/J$ has to be at least 10. The smaller the value of $\Delta v/J$, the greater are the complications due to second-order effects.

Although it is possible, in principle, to analyze a second-order spectrum through computer simulation (Section 9-10), such a spectrum would be much easier to interpret if we could somehow "force" it into a first-order appearance. Since the magnitude of J is determined by the molecule's structure (and is field-independent, Section 8-2), the only way we can increase $\Delta v/J$ is to increase Δv. One way to do this is by using a different spectrometer with a higher field strength and operating frequency (Section 9-9), but this can be a very expensive proposition. In many cases there are much less expensive ways to accomplish the same thing!

A. Solvent-induced shift

Do you recall that the chemical shift of a given nucleus is determined by the strength of the applied magnetic field *as modified by the neighboring electrons and nuclei through shielding effects* (Section 6-1)? These neighboring electrons and nuclei include not only those in the same molecule as the nucleus of interest, but also those in nearby solvent or solute molecules.

For example, the ^1H spectrum of compound **12-1** dissolved in carbon tetrachloride exhibits

12-1

only one sharp signal at δ 1.45 ppm,[1] indicating that the two sets of hydrogens (six methyl and two methylene) are accidentally equivalent (Section 4-4). By contrast, when the ^1H spectrum of **12-1** is examined with benzene (Example 4-3) as solvent, *two* singlets appear [δ 1.11 (2H); 1.17 (6H)], as expected on the basis of symmetry. This effect, known as the **solvent-induced shift** (SIS), is due to the anisotropic field around the benzene molecules (Section 6-6). Because of the way in which the molecules of **12-1** are surrounded by benzene molecules, all eight hydrogens in **12-1** experience extra shielding, but the CH_2 hydrogens are shielded somewhat more than the CH_3 hydrogens.

While the changes in Δv brought about by SIS can be of use in resolving accidental equivalences, they are rarely large enough to turn a second-order spectrum into a first-order one. To do that, we need another trick.

B. Paramagnetic shift reagents

There is a family of compounds which, when added to a solution under nmr examination, can lead to large changes in the chemical shifts of the sample's nuclei. These so-called **paramagnetic shift reagents** (PSR's) are coordination complexes of strongly paramagnetic metal ions of the lanthanide group, such as europium (Eu^{3+}) and praseodymium (Pr^{3+}). One such commercially available shift reagent, known by the abbreviation $Eu(fod)_3$, has structure **12-2**. It is also available in a deuterated version, in which all 30 hydrogens have been replaced with deuterium.

$$F_3C-CF_2-F_2C$$

$$H-C$$

$$(H_3C)_3C$$

$$Eu^{3+}$$

$$Eu(fod)_3$$

12-2

When a PSR is added to a solution containing the sample of interest, there is a weak complexation interaction between the PSR and the sample molecules. The sample molecules are held in relatively close proximity to the metal ion and experience its strong paramagnetic field. As a result, nuclei in the sample molecule usually experience substantial *deshielding*, causing a downfield shift in the position of their signals. The sample nuclei nearest to the metal ion exhibit the greatest shift. Because this is an interaction *through space* (as is NOE, Section 11-3), its magnitude is inversely proportional to r^3, where r is the distance from the metal ion to the nucleus of interest.

Because of the dynamic (Section 12-3) nature of this complexation interaction, the magnitude of PSR-induced shifts is a function of the mole ratio of PSR to sample. Up to a ratio of about 1:1, the greater the ratio, the greater will be the shift. Beyond that ratio, each sample molecule already has its "own" PSR, so additional PSR usually has little effect.

An example of this PSR-induced shift is seen in the 80-MHz ^{1}H spectrum of **12-3**.[2] In the

$$\overset{d}{\underset{\downarrow}{}} \quad \overset{c}{\underset{\downarrow}{}} \quad \overset{b}{\underset{\downarrow}{}} \quad \overset{a}{\underset{\downarrow}{}}$$

$$H_3C-CH_2-CH_2-CH_2-O-CH_2-CH_2-CH_2-CH_3$$

12-3

uncomplexed spectrum (Figure 12-1a), the H_a's show up as the expected triplet (δ 3.40 ppm), the H_d "triplet" (δ 0.91 ppm) is very distorted by second-order effects, while the H_b/H_c multiplet (δ 1.1– 1.8 ppm) is horrible to behold. The addition of deuterated $Eu(fod)_3$ comes to the rescue! It causes the signals of all four sets of hydrogens to shift downfield to varying degrees (Figure 12-1b). When the ratio of $Eu(fod)_3$ to **12-3** reaches 0.72, the chemical shifts are δ_a 5.46 (triplet), δ_b 2.73 (quintet), δ_c 1.91 (sextet with some additional structure), and δ_d 1.11 (triplet). Note how the multiplicity of each signal is now in accord with first-order expectations.

EXAMPLE 12-1   (a) Calculate the magnitude (in ppm) of the PSR-induced shift for each hydrogen in **12-3**. (b) Where in the structure of **12-3** is the metal ion most likely complexed?

Solution: **(a)**

H	PSR-induced shift (ppm)
a	5.46 − 3.40 = 2.06
b	2.73 − 1.45* = 1.28
c	1.91 − 1.45* = 0.46
d	1.11 − 0.91 = 0.20

*Estimated as the center of the multiplet.

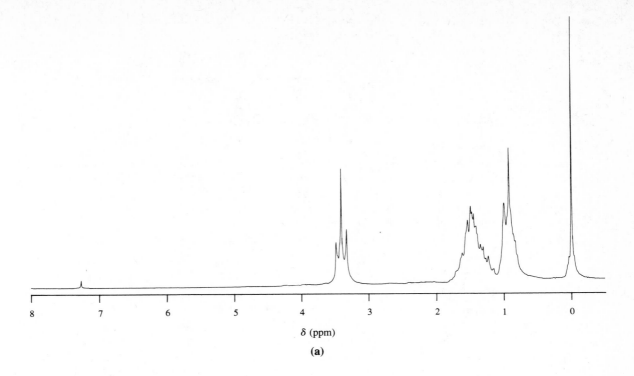

δ (ppm)

(a)

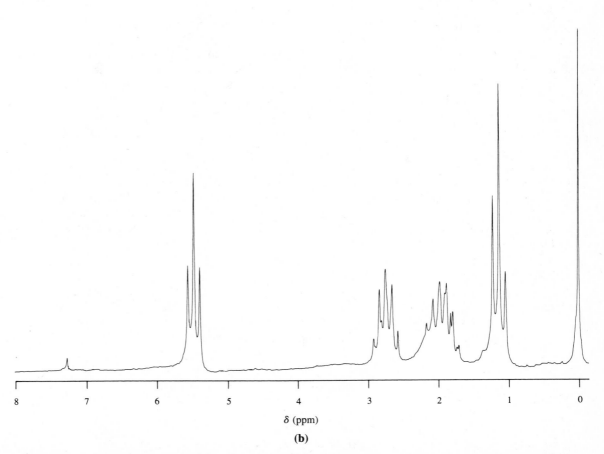

δ (ppm)

(b)

FIGURE 12-1 (a) The 80-MHz ^{1}H spectrum of compound **12-3** dissolved in CDCl$_3$; (b) 80-MHz spectrum of the same sample after addition of 0.72 mole equivalents of deuterated Eu (fod)$_3$.

(b) Because the PSR-induced shifts follow the order $H_a > H_b > H_c > H_d$, complexation of the metal ion must be nearest to H_a, probably at the oxygen atom. Usually, PSR complexation takes place at the most electron-rich site of the sample molecule. This site is often an oxygen, nitrogen, or halogen atom.

12-2. The Study of Dynamic Processes by NMR

A. What is a dynamic process?

It is easy to imagine how useful nmr can be in monitoring the progress of chemical reactions. By following the disappearance of starting material(s) and/or the formation of product(s), we can identify the products, measure rates of reaction, and determine whether reaction intermediates are involved. We've even seen how certain reactions (those involving radical pairs, Chapter 10) provide very special information in their nmr spectra. Such processes, where the structure of the sample changes with time, are described as **dynamic**. But there are other, more subtle, types of dynamic processes that are especially well-suited to nmr investigation.

B. Dynamic physical processes

Molecules are constantly undergoing many types of motion. One type is motion in which the molecule as a whole (including its center of mass) changes position, a motion called **translation**. In addition, there is **rotation** of the molecule around its center of mass, and **vibration** (alternate stretching and compressing) of each bond. There is also rotation around individual bonds that converts one conformation into another (Section 4-2). These types of motion are *physical* processes because they do not involve the breaking of bonds or a change in the sequence (connectivity) of atoms in the molecule. For the most part, these processes are much faster than the nmr time scale (Section 1-7), so an nmr spectrum normally exhibits signals that represent weighted averages over all possible conformations, bond lengths, and positions. We first saw the manifestations of this averaging in Section 4-2 when we examined the hydrogens of the methyl group attached to a benzene ring.

However, we've encountered one instance where rotation around a particular bond in a molecule is restricted, so that the molecule is "locked" in one conformation. The compound is dimethylformamide (**11-2**). Because of a resonance interaction (Sections 7-4 and 9-6) between

11-2

the nitrogen atom and the C=O, there is a certain degree of "double bond character" to the central C—N bond. For this reason, rotation around the bond is restricted and the two methyl groups are thereby rendered *non*equivalent. Another way of saying the same thing is that, at room temperature, not enough thermal energy is available to overcome this resonance interaction.

You are probably aware that as the temperature of a collection of molecules increases, the molecules have more thermal energy, and physical processes such as translation and rotation increase in rate. You might wonder if there is a temperature at which the C—N bond of **11-2** *would* begin to rotate. And if it did, what would be the consequences of such a process?

Rotation of dimethylformamide's C—N bond would bring about *reversible exchange* of the two methyl groups; that is, they would exchange positions between the two possible sites:

At room temperature (25°C) this rotation does not occur to any significant extent (a situation referred to as the **slow exchange limit**), so each methyl group exhibits its own signal (δ 2.98 and 2.81 ppm). However, when dimethylformamide is heated to temperatures above 130° C, rotation becomes fast on the nmr time scale (the so-called **fast exchange limit**), and the two methyl signals are "averaged" to give one sharp singlet at δ 2.895 ppm,[3] midway between the two original signals.

Just as we can calculate the appearance of an nmr spectrum if we know all the chemical shifts and coupling constants (Section 9-10), we can accurately predict the spectrum of a molecule undergoing exchange. The appearance of such a spectrum is determined by the populations (relative amounts) of the various exchanging structures (or sites), the chemical shift difference between the structures *at the slow exchange limit* (Δv_0), and the rate constants (k) for interconversion of the various structures.

For the case of exchange between two equally populated sites (as we have with the methyl groups in dimethylformamide), Eq. (12-1) describes the nmr signal intensity [$I(v)$] as a function of spectral frequency v.[4]

LINE SHAPE FOR TWO-SITE EXCHANGE
$$I(v) = \frac{Ck\,\Delta v_0^2}{2k^2(\bar{v}-v)^2 + 2\pi^2(v_A - v)^2(v_B - v)^2} \tag{12-1}$$

where $\bar{v} = (v_A + v_B)/2$, $\Delta v_0 = v_A - v_B$, and the normalizing constant C has units of Hz. Equation (12-1) is an approximation that neglects the linewidth contribution of T_2. The rate constant k is related to the **lifetime** (τ) of each site (the time that a specific atom occupies the site) by the relation

LIFETIME–RATE CONSTANT RELATION
$$\tau_A = \frac{1}{k_{A \to B}} \quad \text{and} \quad \tau_B = \frac{1}{k_{B \to A}} \tag{12-2}$$

If the two sites are equally populated, both rate constants are equal, as are both lifetimes. Notice also that *slow* processes are ones with *small* rate constants and *long* lifetimes.

Now we can be more explicit about what we mean by the term "nmr time scale." We'll define the **exchange ratio** R by the equation

EXCHANGE RATIO
$$R = \frac{k}{\Delta v_0} \tag{12-3}$$

A *large* value of R indicates *fast* exchange, while a *small* value indicates *slow* exchange. Now let's investigate what happens to the spectrum of dimethylformamide as we increase the exchange ratio from the slow exchange limit to the fast exchange limit.

1. When R is less than ~ 0.1, the system is essentially "frozen" at the slow exchange limit, and Eq. (12-1) predicts two sharp signals, one at v_A, the other at v_B. The halfwidth of these signals is determined by the effective spin–spin relaxation time (Section 3-3): $v_{1/2} = 1/T_2^*$.
2. As the value of R increases (which we can do by increasing the sample's temperature), the signals broaden and move closer together. The separation (Δv) between these broadened signals during slow exchange is given by

PEAK SEPARATION AT SLOW EXCHANGE
$$\Delta v = \Delta v_0 \left[1 - \frac{2R^2}{\pi^2} \right]^{1/2} = \Delta v_0 \left[1 - \frac{2k^2}{\pi^2 \Delta v_0^2} \right]^{1/2} \tag{12-4}$$

EXAMPLE 12-2 (a) Rearrange Eq. (12-4) to express k as a function of Δv and Δv_0. (b) At a certain temperature the 60-MHz ^{1}H spectrum of dimethylformamide exhibits two broad signals with a separation of 9.0 Hz. What are the values of k and R at this temperature?

Solution:

(a)
$$\Delta v = \Delta v_0 \left(1 - \frac{2k^2}{\pi^2 \Delta v_0^2} \right)^{1/2}$$

$$\frac{2k^2}{\pi^2 \Delta v_0^2} = 1 - \frac{\Delta v^2}{\Delta v_0^2} = \frac{\Delta v_0^2 - \Delta v^2}{\Delta v_0^2}$$

$$k^2 = \left(\frac{\pi^2 \Delta v_0^2}{2} \right) \left(\frac{\Delta v_0^2 - \Delta v^2}{\Delta v_0^2} \right) = \frac{\pi^2(\Delta v_0^2 - \Delta v^2)}{2}$$

SLOW-EXCHANGE RATE CONSTANT
$$k = \pi \sqrt{\tfrac{1}{2}(\Delta v_0^2 - \Delta v^2)} \tag{12-5}$$

(b) At 60 MHz, $\Delta v_0 = (2.98 - 2.81)$ ppm (60 Hz ppm^{-1}) = 10.2 Hz. Using Eq. (12-5) gives

$$k = \pi\sqrt{\tfrac{1}{2}(10.2^2 - 9.0^2)} = 10.7 \text{ Hz} = 10.7 \text{ s}^{-1}$$

From Eq. (12-3)

$$R = \frac{k}{\Delta v_0} = \frac{10.7 \text{ Hz}}{10.2 \text{ Hz}} = 1.05$$

---

3. At a certain value of R, the two peaks **coalesce** into single peak centered at $\bar{v}$, but this peak is often so broad it can be difficult to distinguish from background noise.

EXAMPLE 12-3 (a) In terms of Δv_0, what are the values of k and R at coalescence? (b) At what value of k will the methyl signals in the 60-MHz ^{1}H spectrum of dimethylformamide coalesce?

Solution:

(a) By the definition of coalescence, Δv must equal zero. Substituting zero for Δv in Eq. (12-5), we find

COALESCENCE RATE CONSTANT
$$k_c = \pi\sqrt{\frac{\Delta v_0^2}{2}} = \frac{\pi}{\sqrt{2}}\Delta v_0 = 2.22(\Delta v_0) \tag{12-6}$$

and therefore

$$R_c = \frac{k_c}{\Delta v_0} = \frac{2.22\,\Delta v_0}{\Delta v_0} = 2.22$$

(b) Using Eq. (12-6) and our value of Δv_0 from Example 12-2b, we find

$$k_c = 2.22(10.2 \text{ Hz}) = 22.6 \text{ Hz} = 22.6 \text{ s}^{-1}$$

---

4. As the rate of exchange increases still further, the single peak at $\bar{v}$ begins to sharpen. Its halfwidth ($v_{1/2}$) is given by

MODERATELY FAST EXCHANGE HALFWIDTH
$$v_{1/2} = v^{\circ}_{1/2} + \frac{\pi \Delta v_0^2}{2k} \tag{12-7}$$

where $v^{\circ}_{1/2}$ is the additional line width due to spin–spin relaxation[5] $v^{\circ}_{1/2} = 1/T_2^*$ and is usually taken to be equal to the halfwidth of the signal at the fast exchange limit.

EXAMPLE 12-4 (a) Solve Eq. (12-7) for k. (b) What are the values of k and R for the methyl exchange in dimethylformamide if $v_{1/2} = 3.1$ Hz and $v^{\circ}_{1/2} = 1.0$ Hz?

Solution:

(a)

$$v_{1/2} = v^{\circ}_{1/2} + \frac{\pi \Delta v_0^2}{2k} \qquad v_{1/2} - v^{\circ}_{1/2} = \frac{\pi \Delta v_0^2}{2k}$$

FAST EXCHANGE RATE CONSTANT
$$k = \frac{\pi \Delta v_0^2}{2(v_{1/2} - v^{\circ}_{1/2})} \tag{12-8}$$

(b) Applying Eq. (12-8) gives us

$$k = \frac{\pi(10.2 \text{ Hz})^2}{2(3.1 - 1.0) \text{ Hz}} = 77.8 \text{ Hz} = 77.8 \text{ s}^{-1}$$

and

$$R = \frac{k}{\Delta v_0} = \frac{77.8 \text{ Hz}}{10.2 \text{ Hz}} = 7.63$$

5. Finally, when R is greater than ~ 30, the system is essentially at the fast exchange limit, and the spectrum consists of one sharp signal at $\bar{v}$ with $v_{1/2} = v^{\circ}_{1/2}$.

These five conditions are depicted graphically in Figure 12-2, which was generated from Eq. (12-1) with the values of the rate constants calculated above and the following data: $v_A = 178.8$ Hz, $v_B = 168.6$ Hz, $\Delta v_0 = 10.2$ Hz, $\bar{v} = 173.7$ Hz, and $C = 0.4195$ Hz.

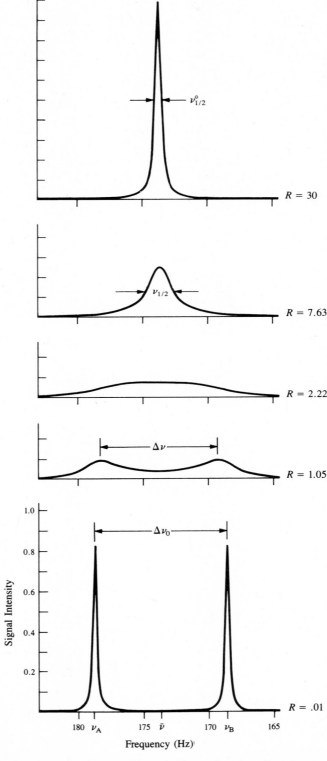

FIGURE 12-2 Line shape for two-site exchange [Eq. (12-1)] as a function of exchange ratio R. Values are from the 60-MHz ^{1}H spectrum of dimethylformamide.

Whenever a spectrum shows *reversible* changes as the sample temperature is varied, one should suspect that dynamic processes are at work. In a full investigation of a dynamic process by nmr, the first step is to determine the values of Δv_0 and the fractions of the molecular population in each conformation or site at the slow exchange limit. Next, the value of $v_{1/2}^{\circ}$ is measured at the fast exchange limit. Then several additional spectra are recorded at various temperatures between these limits, including one near coalescence. Finally, the rate constant values for each of the intermediate spectra are calculated from equations such as (12-1) through (12-8). For systems that involve more than simple two-site exchange, the calculations usually require a computer, and programs for this purpose are widely available. One such program is DNMR3 by D. A. Kleier and G. Binsch and is available from the Quantum Chemistry Program Exchange at Indiana University, Bloomington, IN 47401. The practice of matching an actual spectrum with a computer-generated spectrum is referred to as **complete lineshape analysis**.

Figure 12-3 is a complete computer simulation of a coupled AB spin system (Section 9-11) undergoing exchange. An actual example of this type of system is the "ring flipping" of a cyclohexane derivative (structure **12-4**). Notice how this process exchanges the sites (environments) of H_a and H_b though the two structures **i** and **ii** are otherwise equivalent. Values used in creating this figure were $v_a = 120$ Hz, $v_b = 60$ Hz, $J_{ab} = 10$ Hz, and $v_{1/2}^{\circ} = 2$ Hz. The spectra are plotted as a function of τ_i, the lifetime of structure **12-4i** ($\tau_i = 1/k_{i \rightarrow ii}$). Don't confuse this figure with the 2DNMR spectra we discussed in Section 11-9 just because they too are displayed in a three-dimensional perspective.

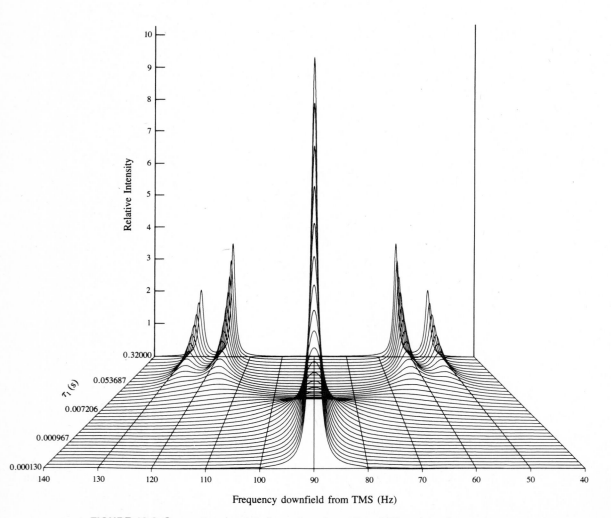

FIGURE 12-3 Computer-simulated spectra of an AB $\rightleftarrows$ A′B′ exchanging system.

12-4i   **12-4ii**

EXAMPLE 12-5 From the data used to generate Figure 12-3, calculate Δv_0 and τ at coalescence.

Solution:

$$\Delta v_0 = v_a - v_b = (120 - 60) \text{ Hz} = 60 \text{ Hz}$$

Find the value of the rate constant at coalescence from Eq. (12-6), and substitute it into Eq. (12-2):

$$\tau_i = \frac{1}{k_{i \to ii}} = \frac{1}{(2.22 \, \Delta v_0)} = \frac{1}{2.22(60 \text{ Hz})} = 0.0075 \text{ s}$$

Note in Figure 12-3 that at this value of τ_i the spectrum consists of one very broad signal.

EXAMPLE 12-6 Suppose you are investigating the dynamic behavior of an exchange process involving the two noncoupled methyl groups in dimethylformamide. At the slow exchange limit the 60-MHz ^{1}H spectrum shows singlets for the methyl *hydrogens* at δ 2.98 and 2.81 ppm, while the 20-MHz ^{13}C spectrum shows signals for the methyl *carbons* at δ 36.2 and 31.1 ppm. **(a)** What is the value of Δv_0 for both spectra? **(b)** Which spectrum will exhibit coalescence at the higher temperature?

Solution:

(a) For the hydrogens,

$$\Delta v_0 = (2.98 - 2.81) \text{ ppm } (60 \text{ Hz ppm}^{-1}) = 10.2 \text{ Hz}$$

and for the carbons

$$\Delta v_0 = (36.2 - 31.1) \text{ ppm } (20 \text{ Hz ppm}^{-1}) = 102 \text{ Hz}$$

(b) Because coalescence [Eq. (12-6)] requires that $k = 2.22(\Delta v_0)$, the carbons require a higher exchange rate constant (and hence a higher temperature) to make their signals coalesce.

C. Dynamic chemical processes

Many molecules undergo reversible *chemical* changes, that is, reorganization of bonds and changes in bonding sequence. Such molecules are called **fluxional molecules**. As with dynamic physical processes, these dynamic chemical processes are also particularly amenable to study by nmr.

One classic example is the molecule bullvalene, structure **12-5**.[5] This unique structure has a

12-5

C_3 axis (Section 4-1), and its hydrogens constitute an AA′A″BB′B″CC′C″D spin system. At low temperature (below $-85°$ C) nothing extraordinary occurs, and the 60-MHz ^{1}H spectrum shows two very complex multiplets, one at δ 2.2 ppm (H_a, H_d) and the other at δ 5.6 ppm (H_b, H_c). But as the temperature of the sample is raised, the two multiplets broaden and move together. At 15°C the spectrum consists of one extremely broad signal ($v_{1/2} = 150$ Hz). And at 120°C there is but *one* sharp singlet!

EXAMPLE 12-7 Predict the chemical shift of the bullvalene signal at the fast exchange limit.

Solution: Remember that this signal is the weighted average of all exchanging signals. The spectrum of bullvalene at the slow exchange limit has a 6H multiplet at δ 5.6 ppm and a 4H multiplet at δ 2.2 ppm. Thus, the average chemical shift of all 10 hydrogens is

$$\delta_{av} = \frac{6(5.6 \text{ ppm}) + 4(2.2 \text{ ppm})}{10} = 4.2 \text{ ppm}$$

EXAMPLE 12-8 Would the signals in the 300-MHz ^{1}H spectrum of bullvalene coalesce at the same, higher, or lower temperature than the signals in the 60-MHz spectrum?

Solution: Recall that Δv, like v, increases linearly with field strength and spectrometer frequency. So

at 60 MHz: $\Delta v_0 = (5.6 - 2.2) \text{ ppm} (60 \text{ Hz ppm}^{-1}) = 204 \text{ Hz}$
at 300 MHz: $\Delta v_0 = (5.6 - 2.2) \text{ ppm} (300 \text{ Hz ppm}^{-1}) = 1020 \text{ Hz}$

Thus, from Eq. (12-6), we see that coalescence requires an exchange rate five times greater (hence, a *higher* temperature) at 300 MHz than at 60 MHz.

The way in which all ten hydrogens in bullvalene become equivalent involves an almost infinite series of **degenerate** (i.e., leading to an equivalent species) **rearrangements** such as

Many examples of this type of behavior are now known. But had it not been for nmr, most would never have been discovered.

12-3. CPMAS Solid-Phase NMR

Although we mentioned the fact in Section 3-4C, we have not often emphasized that the vast majority of nmr spectra involve samples in the liquid state, either pure liquids or solutions. But is it possible to obtain nmr spectra on solid samples? And why might we want to?

There are two main reasons for examining solid samples by nmr. The first is trivial: perhaps no solvent can be found that will dissolve the sample! But consider also the situation where we wish to investigate the nature of the sample *in the solid state*, to study crystal structures and interactions that change or disappear when the sample is dissolved.

All the nmr spectra we've looked at so far in this book are considered **high-resolution spectra** because typical halfwidths of lines (except in cases involving exchange, Section 12-2) are far smaller than the chemical shift differences and coupling constants that we use to elucidate the structure of the sample. Until the early 1980s, attempts to examine solid-state samples by nmr usually gave **low-resolution spectra** with such horrendously broad lines that little useful information could be derived from them.

There are several reasons why solid samples give spectral lines that are so much broader than those from liquid samples. In solution, dynamic processes such as translation and rotation serve to "average out" most local anisotropic effects (Section 12-2). In solids, however, these processes are greatly retarded or stopped altogether, so their "averaging" effects are lost. Thus, in solid samples there is a range of local environments for each set of nuclei and a range of orientations with respect to the applied magnetic field, leading to a range of chemical shifts (rather than a single value) for each set and hence to broad lines. This effect is referred to as **chemical-shift anisotropy** (CSA). Signals are also broadened by unaveraged magnetic dipole–dipole interactions through space (Section 11-3) and, for nuclei with $I > \frac{1}{2}$, electric quadrupole interactions. What's more, spin–lattice relaxation (Section 2-3B) is very inefficient in solids, leading to long T_1 values. This fact means that long relaxation delays are required when the PFT method (Section 3-3) is used on solid samples. Fortunately, a technique has been developed that circumvents most of these problems by artificially "averaging out" many of the anisotropies while also overcoming the effects of slow relaxation. It is called the **cross-polarization**, **magic angle spinning** (CPMAS) technique.[6]

The energy E of interaction between two magnetic dipoles μ is described by

MAGNETIC DIPOLE INTERACTION
$$E = \frac{\mu_1\mu_2(1 - 3\cos^2\theta)}{r^3}$$
(12-9)

where θ is the angle the pair of dipoles makes with the applied magnetic field. In *liquid* samples this energy is averaged to zero by the rapid rotation of the molecules. However, rapidly spinning a *solid* sample around an axis that makes a certain "magic angle" with the applied magnetic field also averages this interaction energy to zero, causing the spectral lines to be sharpened considerably.

EXAMPLE 12-9 What is the value of the "magic angle"?

Solution: Set E equal to zero in Eq. (12-9) and solve for θ.

$$E = 0 = \frac{\mu_1\mu_2(1 - 3\cos^2\theta)}{r^3}$$

$$1 = 3\cos^2\theta$$

$$\theta = \cos^{-1}(\sqrt{1/3}) = 54° \, 44'$$

The rate of rotation around this magic axis must be greater than the frequency of the dipole–dipole interaction; spin rates of 4000–5000 Hz are typical.

To overcome the effects of slow relaxation, we can use cross-polarization techniques (like those mentioned in Sections 11-3 and 11-9) if the sample is suitably constituted. For example, cross polarization can be especially useful for generating ^{13}C spectra of compounds that possess many hydrogens. The abundant hydrogens are irradiated with a powerful decoupling field, and their resulting polarization is transferred to the nearby, but less abundant, ^{13}C nuclei (as in the HETCOR 2DNMR technique, Section 11-9). The effectiveness of this technique depends on the fact that the hydrogens relax much faster than the carbons, so the long relaxation delays between pulses are circumvented.

For powdered (finely ground) crystalline solids, the spectra generated by CPMAS enjoy quite decent resolution, with halfwidths of a few hertz. Figure 12-4 is the CPMAS ^{13}C spectrum of the highly insoluble compound **12-6**.[7] Although not all of the CH_2 groups are fully resolved, the simplicity of the spectrum and the sharp signals at δ 217.6, 83.5, and 52.5 ppm constitute convincing evidence for the

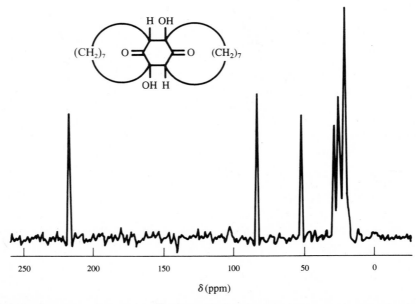

FIGURE 12-4 The solid-phase ^{13}C CPMAS spectrum (25.18 MHz) of compound **12-6**.

symmetrical structure shown. (See Problem 6 in Self-Test II.)

$$H_2C-H_2C-H_2C$$

(structure 12-6, a symmetrical cyclic molecule)

H OH

$H_2C-H_2C-H_2C\diagdown$ $\underset{C}{\overset{|}{C}}-\underset{C}{\overset{|}{C}}\diagup CH_2-CH_2-CH_2$

H_2C $O{=}C$ $C{=}O$ CH_2

$H_2\overset{}{C}-H_2C-H_2C\diagup \underset{C}{C}-\underset{C}{C}\diagdown CH_2-CH_2-CH_2$

HO H

12-6

The field of solid-phase nmr has expanded rapidly to include the investigation of polymers, liquid crystals, inorganic catalysts, coal, and a host of other materials that were previously opaque to nmr.

12-4. Surface Coil NMR and Magnetic Resonance Imaging

By now, we have used nmr successfully to obtain detailed information about our samples. We've been able to deduce the structure of unknown molecules, monitor chemical reactions and other dynamic processes, and even investigate the properties of solid materials. But some of the most exciting new applications of nmr are in the fields of biochemistry, biology, and medicine.

A. Surface coil nmr

It is now possible, for example, to study metabolic processes within a *living* animal by nmr techniques.[8] The anesthetized animal is confined to a strong magnetic field (~ 4 T), and a small transmitter/receiver coil (the so-called **surface coil**) is placed at the desired location. Data are collected in the usual way (Chapter 3) to provide spectra of the chosen nucleus near the surface coil. Phosphorus is especially important to biochemists because it is part of so many metabolically strategic compounds (e.g., ATP, ADP, RNA, and DNA). A quick review of Table 2-1 shows that ^{31}P, because of its 100% natural abundance and decent sensitivity, is an ideal isotope for nmr study.

EXAMPLE 12-10 At what frequency does ^{31}P undergo resonance in a 4.28-T magnetic field?

Solution: Use Eq. (2-6) and the value of γ for ^{31}P from Table 2-1.

$$v = \frac{\gamma B_0}{2\pi} = \frac{(108.29 \times 10^6 \text{ rad T}^{-1}\text{s}^{-1})(4.28 \text{ T})}{2\pi \text{ rad}} = 73.8 \text{ MHz}$$

The beauty of the surface coil technique is that it is noninvasive and exposes the animal to only a magnetic field and rf radiation, neither of which has any known deleterious effects.

B. Magnetic resonance imaging

"Magnetic Fields Send Physicians a Clear Signal for Treating Patients" proclaims the title of a recent article in the *Chicago Tribune*.[9] Indeed, nmr has started something of a revolution in diagnostic medicine with the advent of a technique known as **magnetic resonance imaging** (MRI).[10]

Almost every one of us, at one time or another, has had a diagnostic X-ray taken. The ultimate modern version of this tool is a technique known as CAT (computerized axial tomography) scanning, wherein the patient is subjected to a series of X-ray "snapshots" taken from various positions and angles around the anatomical region of interest. A computer collects these data and generates a three-dimensional composite image of the suspect body part. The risk of such a procedure, of course, is that X-rays are beams of extremely energetic electromagnetic radiation with $\sim 10^{12}$ times the energy of rf radiation (Table 1-1). As a result, even relatively low levels of exposure to X-rays are a source of concern. (Why do you think X-ray technicians always wear those lead-lined aprons?)

MRI instruments currently available permit low-resolution nmr examination of living subjects. The magnets used in such instruments are generally superconducting magnets that generate field strengths on the order of 0.15–0.6 T. The sampling volume is large enough to accommodate an entire human body. As in surface coil nmr, the transmitter/receiver coil is positioned outside the body. ^1H is usually the nucleus chosen for measurement (because of its high abundance and sensitivity), though other nuclei can also be investigated. By means of a technique known as **planar imaging**, the body part of interest is examined region by region from various angles, then a computer reconstructs the resulting data into a three-dimensional map (image) of the ^1H density in that body region. Because each type of living tissue has a perceptibly different ^1H density with different relaxation characteristics, each gives rise to signals of different intensity. The resulting images, when viewed as two-dimensional "slices," resemble conventional CAT scans. In principle, however, more kinds of information, such as the nature of any metabolic disorders, are available from the nmr data. Most importantly, though, MRI is noninvasive, completely painless, requires no radioactive or radio-opaque compounds to be taken internally and, above all, avoids exposure to the hazards of X-rays.

12-5. Afterword

Well, we've come a long way together, from the basic physical properties of electromagnetic radiation to the most contemporary aspects of nmr. I hope you enjoyed the journey. If you have digested this material successfully, you now possess a quite sophisticated knowledge about what nmr is, and what it can do. I sincerely hope this information will prove useful to you in your career.

Self-Test II presents a number of problems that review most of the concepts we've covered. Take time to try them. The answers are in Appendix 2, with references to the sections where the particular topic is covered. Then, from time to time, open the book and work a problem or two for old time's sake. This will help lubricate those synaptic junctions in your brain so you won't lose that which you've worked so hard to gain. Good luck!

REFERENCES

1. K. C. Lilje and R. S. Macomber, *J. Org. Chem.* **39**, 3600 (1974).
2. R. E. Rondeau and R. E. Sievers, *J. Am. Chem. Soc.* **93**, 1522 (1971).
3. M. Rabinowitz and A. Pines, *J. Am. Chem. Soc.* **91**, 1585 (1969).
4. E. D. Becker, *High Resolution NMR*, 2nd ed. Academic Press, New York, 1980.
5. P. Laszlo and P. Stang, *Organic Spectroscopy*. Harper and Row, New York, 1971.
6. G. E. Maciel, *Science* **226**, 282 (1984).
7. R. S. Macomber and P. Murphy, *Organic Magnetic Resonance* **22**, 255 (1984).
8. L. W. Jelinski, *Chem. Eng. News* **62**, 26 (1984).
9. J. Van, *Chicago Tribune*, November 30, 1986, Section 6, P-1.
10. E. R. Andrews, I. R. Young, and G. M. Bydder, in *NMR in the Life Sciences* (E. M. Bradbury and C. Nicolini, eds.). Plenum, New York, 1986; S. R. Thomas and R. L. Dixon, *Physics News in 1986* (a special report in *Physics Today*), **40**, S-42 (1987).

SUMMARY

1. The chemical shifts of nuclei within solute molecules in solution are influenced to some extent by the nature of the solvent. In many cases, changing the solvent causes a significant shift (called the solvent-induced shift) in the position of the solute's signals.
2. There exists a sizeable family of lanthanide-group metal complexes, known as paramagnetic shift reagents (PSRs), which cause often dramatic downfield shifts in the signals of a solute. The closer a solute molecule's nuclei are to the paramagnetic metal ion, the larger the downfield shift in their signal. The main use of PSRs is to separate (resolve) complex signals, making the spectrum more "first-order" in appearance.
3. Nmr spectroscopy is a very useful tool for the study of dynamic molecular processes, i.e., processes that cause the sample to change with time. There are two basic types of dynamic processes, reversible and irreversible. An example of an irreversible process is a chemical reaction that converts reactant to product, A → B. A reversible process, denoted A ⇌ B, involves the interconversion of two (or

more) structures by either chemical changes (i.e., breaking and making of bonds), or physical changes (e.g., rotations around bonds). The appearance of a dynamic nmr spectrum is highly dependent on the rates of these processes.

4. The time scale of a dynamic nmr measurement can be expressed in terms of an exchange ratio $R (= k/\Delta v_0)$. For the simplest dynamic case, that of two-site exchange, the value of R determines the appearance of the spectrum:

   (a) The slow exchange limit ($R < 0.1$): two sharp lines, one at v_A, the other at v_B.
   (b) Moderately slow exchange ($0.1 < R < 2$): two broadened signals separated by

   $$\Delta v = \Delta v_0 \left[1 - \frac{2R^2}{\pi^2} \right]^{1/2} \qquad \text{(12-4)}$$

   (c) Coalescence ($R = 2.22$): one extremely broad signal.
   (d) Moderately fast exchange ($2.5 < R < 30$): one broadened signal with halfwidth given by

   $$v_{1/2} = v^\circ_{1/2} + \frac{\pi \Delta v_0^2}{2k} \qquad \text{(12.7)}$$

   (e) The fast exchange limit ($R > 30$): One very sharp line with halfwidth $v^\circ_{1/2}$, located at $(v_A + v_B)/2$.

6. The procedure for determining the rates of reversible dynamic processes by computer analysis of nmr spectra is termed complete lineshape analysis.

7. The cross-polarization, magic angle spinning (CPMAS) technique is used to acquire nmr spectra of solid (microcrystalline) samples. Cross polarization (e.g., $^{13}C\{^1H\}$) helps overcome sensitivity problems normally associated with very slow spin–lattice relaxation in solids. Magic angle (54°44′) spinning serves to average out the magnetic dipole interactions of molecules in the solid state, giving much sharper lines than observed when the sample is spun at any other angle to the applied magnetic field.

8. Nmr measurements on living organisms (including human subjects) can be accomplished by placing the organism in a strong magnetic field and locating the probe at the surface near the area of interest. Alternatively, if the probe is scanned over a region of interest, the resulting data can be reconstructed by a computer into a three-dimensional map of the population density of the target nucleus. Such magnetic resonance imaging (MRI) techniques provide images of living tissue that resemble but are in fact complementary to conventional CAT scans. MRI "pictures" are of immense diagnostic value to the physician, and MRI completely circumvents the hazards of X-ray exposure.

SELF-TEST II

The following ten multi-part questions range in difficulty from relatively straightforward to quite difficult. The answers are in Appendix 2, but don't peek at them until you've given each problem your best shot!

Each of the first three problems consists of a molecular formula, a ^1H spectrum, and a $^{13}C\{^1H\}$ spectrum of an unknown compound. For each of these three sets of spectra,

(a) Convert the spectral data to condensed form.
(b) Propose a molecular structure consistent with both spectra.
(c) Assign all signals in both spectra to nuclei in your structure by calculating the expected position of each resonance.
(d) Answer any additional questions posed in the problem.

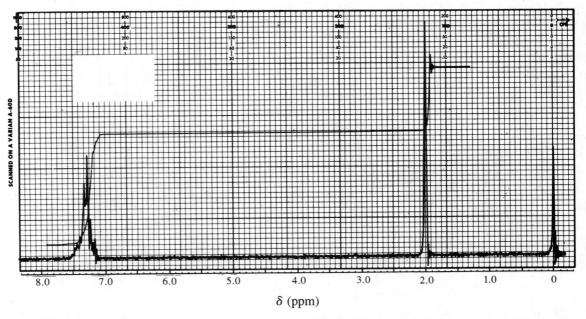

FIGURE II-1 The 60-MHz ^1H spectrum of C_9H_8. © Sadtler Research Laboratories, Division of Bio-Rad Laboratories, Inc., 1972.

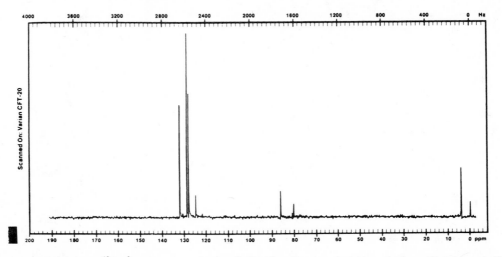

FIGURE II-2 The 20-MHz ^{13}C spectrum of C_9H_8. © Sadtler Research Laboratories, Division of Bio-Rad Laboratories, Inc., 1983.

1. Figures II-1 and II-2 are the 60-MHz ^{1}H and 20-MHz ^{13}C spectra of a compound with the molecular formula C_9H_8.
   (d) Predict the appearance of the off-resonance decoupled ^{13}C spectrum of this compound, with ν_2 set 2000 Hz away from the average ^{1}H chemical shift and B_2 equal to 2×10^{-4} T.

2. Figures II-3 and II-4 are the 60-MHz ^{1}H and 20-MHz ^{13}C spectra of a compound with the molecular formula $C_7H_{14}O_2$.
   (d) Predict the appearance of the APT ^{13}C spectrum of this compound with τ set to 8 ms.

3. Figures II-5 and II-6 are the 60-MHz ^{1}H and 20-MHz ^{13}C spectra of a compound with the molecular formula $C_7H_6ClNO_2$.
   (d) By drawing graphs similar to Figure 11-15, predict the appearance of the HETCOR 2D NMR spectrum of this compound, first with $t_1 = 4.0$ ms, then with $t_1 = 3.1$ ms.

4. Predict the position and intensity of each line in the *coupled* ^{13}C spectra of

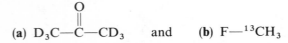

(a) $D_3C-\overset{\displaystyle O}{\overset{\|}{C}}-CD_3$ and   (b) $F-^{13}CH_3$

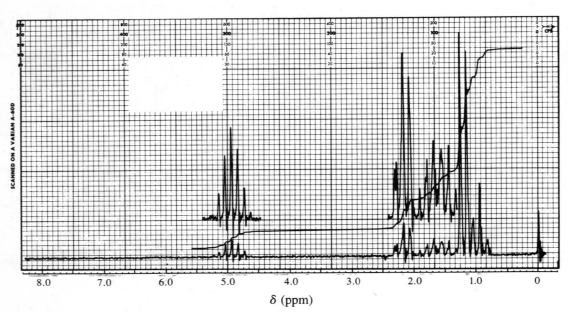

FIGURE II-3 The 60-MHz ^{1}H spectrum of $C_7H_{14}O_2$. © Sadtler Research Laboratories, Division of Bio-Rad Laboratories, Inc., 1976.

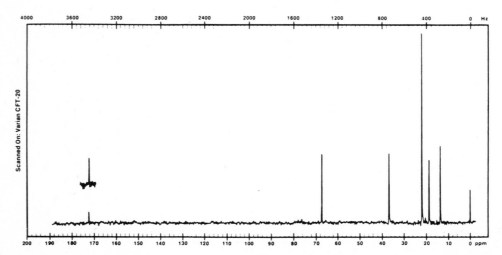

FIGURE II-4 The 20-MHz ^{13}C spectrum of $C_7H_{14}O_2$. © Sadtler Research Laboratories, Division of Bio-Rad Laboratories, Inc., 1983.

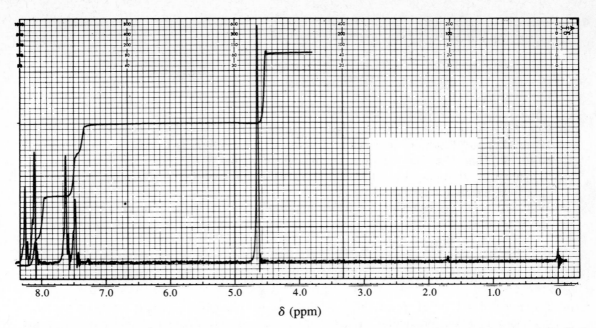

FIGURE II-5 The 60-MHz ^1H spectrum of $C_7H_6ClNO_2$. © Sadtler Research Laboratories, Division of Bio-Rad Laboratories, Inc., 1967.

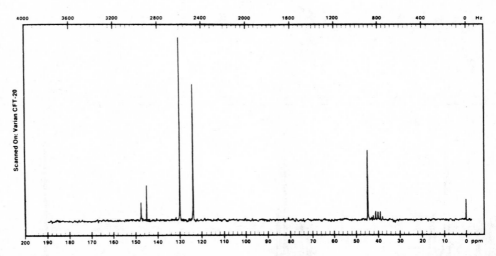

FIGURE II-6 The 20-MHz ^{13}C spectrum of $C_7H_6ClNO_2$. © Sadtler Research Laboratories, Division of Bio-Rad Laboratories, Inc., 1983.

5. Cyclooctatetraene exhibits only one signal in its ^1H spectrum, at δ 5.75 ppm. Compare this value to the chemical shift of benzene (Section 6-6) and discuss its significance.

cyclooctatetraene   benzene

6. (a) Cyclohexanone exhibits ^{13}C signals at δ 25.1, 27.2, 41.9, and 211.2 ppm. Assign these signals to the appropriate carbons in the following structure:

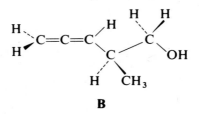

cyclohexanone

(b) Refer back to the CPMAS ^{13}C spectrum of structure **12-6** (Figure 12-4). Assign the signals at δ 217.6, 83.5, and 52.5 ppm to specific carbons in the structure by calculating their expected positions. [*Hint:* Use cyclohexanone as a model compound.]

(c) By symmetry, how many sets of CH_2 carbons are there in structure **12-6**? Can you assign any of these to signals in the spectrum?

7. (a) Assuming first-order behavior, predict the appearance of the ^1H spectrum of compound **A**.

$$\underset{H}{\overset{H}{>}}C=C=C\underset{H}{\overset{C(CH_3)_3}{<}}$$

A

(b) The actual 60-MHz ^1H spectrum of **A** is shown in Figure II-7. Account for any differences between it and your predictions. What are the small signals flanking the singlet at δ 1.0 ppm?

(c) Next, consider the related molecule **B**. By labeling each hydrogen, indicate which, if any, are symmetry-equivalent. You can assume free rotation around all single bonds.

B

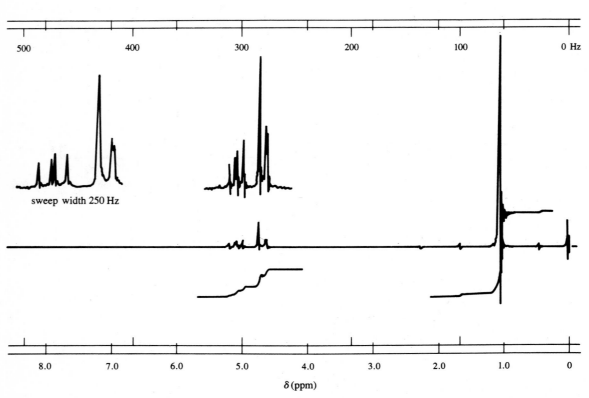

FIGURE II-7 The 60-MHz ^1H spectrum of compound **A**.

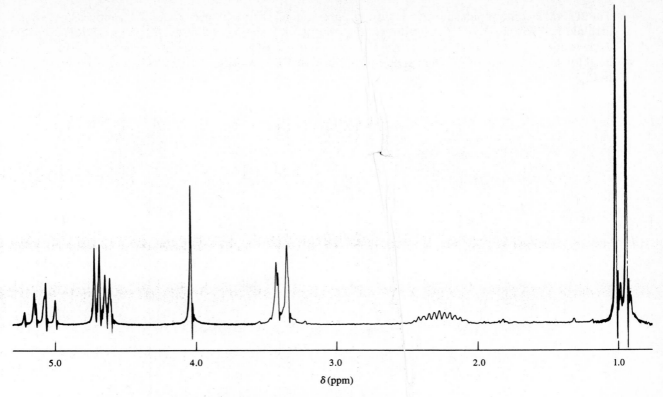

FIGURE II-8 The 90-MHz ^1H spectrum of compound **B**.

(d) Section 6-7 states that O—H and N—H resonances are broadened and decoupled by exchange processes. Explain how this exchange broadens the signals.

(e) How many δ and J values would be needed to generate an accurate computer simulation of the ^1H spectrum of **B**? Estimate the magnitude of each of these quantities.

(f) Figure II-8 is the 90-MHz ^1H spectrum of **B**. Assign all signals. According to first-order rules, what is the maximum number of lines in the multiplet at δ 2.3 ppm?

(g) Predict the result on the rest of the spectrum in Figure II-8 of irradiating the signal at δ 2.3 ppm.

(h) What is unique about the ^{13}C signal from the middle carbon in a C=C=C linkage?

8. Consider structures **C** and **D**. The ^1H spectrum of **C** is *independent* of temperature, showing (among other signals) *two* singlets for the OCH_3 groups [δ 4.04 (6H), 4.10 (3H)]. By contrast, the ^1H spectrum of **D** *is* reversibly temperature-dependent. At +38°C there are *two* OCH_3 singlets [δ 3.44 (6H), 3.79 (3H)], at −23°C only *one* is observed (δ 3.79), and at −67°C there are *three* [δ 2.76 (3H), 3.79 (3H), and 4.13 (3H)]. Account for these observations.

C

D

9. The 2D NMR contour plot shown in Figure 11-13b is actually a COSY spectrum of the complicated alkaloid E. The 300-MHz ^1H spectrum of this compound is shown in Figure II-9 (the signal at δ 7.259 is due to $CHCl_3$, an impurity in the solvent). Figure II-10 is a combination of Figures 11-13b and II-9. Use Figure II-10 to assign as many signals in the ^1H spectrum as you can to the hydrogens in E.

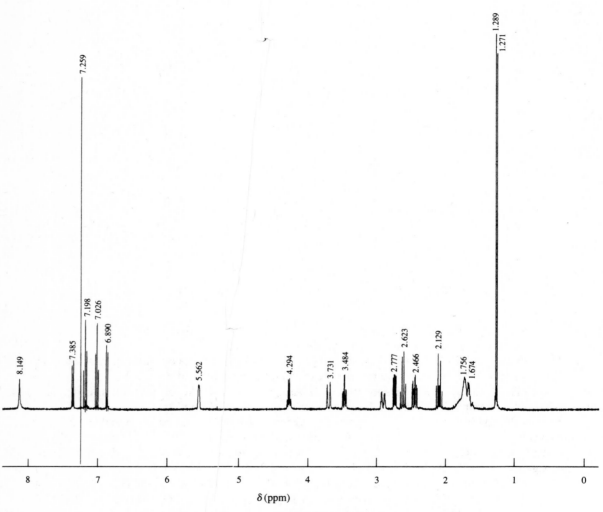

FIGURE II-9 The 300-MHz ^1H spectrum of compound **E**.

E

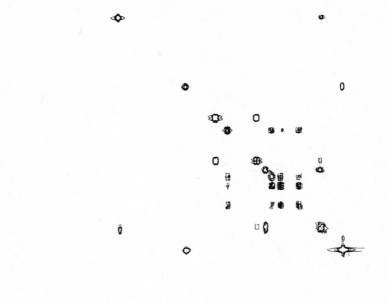

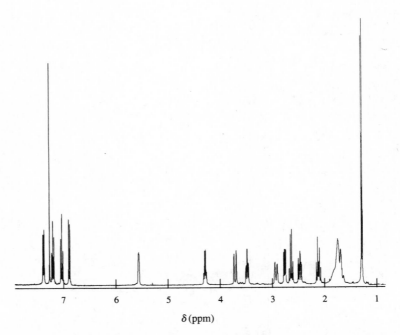

FIGURE II-10 Combined 300-MHz ^{1}H spectrum and 2D COSY spectrum of compound **E**.

10. In Example 2-2 (and Table 2-1) we encountered the two isotopes of chlorine. Each of these isotopes has an *I* value of $\frac{3}{2}$ and a substantial γ value, as well as an electric quadrupole moment. Yet in all the nmr spectra of chlorine-containing compounds we've discussed there has never been any evidence of coupling between the chlorine and the other hydrogens and carbons in the molecules. In contrast, hyperfine coupling between the chlorine and the unpaired electron is commonly observed in the esr spectra of chlorine-containing radicals. Why is coupling to chlorine observed in esr spectra but not in nmr spectra?

APPENDIX 1: Periodic Table of the Elements

s Orbitals being filled

d Orbitals being filled

p Orbitals being filled

Noble gases

Transition elements

Period number = n, the highest occupied electron level.

Group numbers

Period	IA ns^1	IIA ns^2	IIIB $(n-1)d^1ns^2$	IVB $(n-1)d^2ns^2$	VB $(n-1)d^3ns^2$	VIB $(n-1)d^4ns^2$	VIIB $(n-1)d^5ns^2$		VIIIB $(n-1)d^6ns^2$, $(n-1)d^7ns^2$, $(n-1)d^8ns^2$			IB $(n-1)d^{10}ns^1$	IIB $(n-1)d^{10}ns^2$	IIIA ns^2np^1	IVA ns^2np^2	VA ns^2np^3	VIA ns^2np^4	VIIA ns^2np^5	VIIIA ns^2np^6
1	**H** 1 $1s^1$ 1.0079																		**He** 2 $1s^2$ 4.0026
2	**Li** 3 $2s^1$ 6.941	**Be** 4 $2s^2$ 9.01218												**B** 5 $2s^22p^1$ 10.81	**C** 6 $2s^22p^2$ 12.011	**N** 7 $2s^22p^3$ 14.0067	**O** 8 $2s^22p^4$ 15.9994	**F** 9 $2s^22p^5$ 18.9984	**Ne** 10 $2s^22p^6$ 20.179
3	**Na** 11 $3s^1$ 22.9898	**Mg** 12 $3s^2$ 24.305												**Al** 13 $3s^23p^1$ 26.9815	**Si** 14 $3s^23p^2$ 28.086	**P** 15 $3s^23p^3$ 30.9738	**S** 16 $3s^23p^4$ 32.06	**Cl** 17 $3s^23p^5$ 35.453	**Ar** 18 $3s^23p^6$ 39.948
4	**K** 19 $4s^1$ 39.098	**Ca** 20 $4s^2$ 40.08	**Sc** 21 $3d^14s^2$ 44.959	**Ti** 22 $3d^24s^2$ 47.90	**V** 23 $3d^34s^2$ 50.9414	**Cr** 24 $3d^54s^1$ 51.996	**Mn** 25 $3d^54s^2$ 54.938	**Fe** 26 $3d^64s^2$ 55.847	**Co** 27 $3d^74s^2$ 58.9332	**Ni** 28 $3d^84s^2$ 58.70	**Cu** 29 $3d^{10}4s^1$ 63.546	**Zn** 30 $3d^{10}4s^2$ 65.38	**Ga** 31 $4s^24p^1$ 69.72	**Ge** 32 $4s^24p^2$ 72.59	**As** 33 $4s^24p^3$ 74.9216	**Se** 34 $4s^24p^4$ 78.96	**Br** 35 $4s^24p^5$ 79.904	**Kr** 36 $4s^24p^6$ 83.80
5	**Rb** 37 $5s^1$ 85.4678	**Sr** 38 $5s^2$ 87.62	**Y** 39 $4d^15s^2$ 88.9059	**Zr** 40 $4d^25s^2$ 91.22	**Nb** 41 $4d^45s^1$ 92.9064	**Mo** 42 $4d^55s^1$ 95.94	**Tc** 43 $4d^55s^2$ (97)	**Ru** 44 $4d^75s^1$ 101.07	**Rh** 45 $4d^85s^1$ 102.905	**Pd** 46 $4d^{10}$ 106.4	**Ag** 47 $4d^{10}5s^1$ 107.868	**Cd** 48 $4d^{10}5s^2$ 112.40	**In** 49 $5s^25p^1$ 114.82	**Sn** 50 $5s^25p^2$ 118.69	**Sb** 51 $5s^25p^3$ 121.75	**Te** 52 $5s^25p^4$ 127.60	**I** 53 $5s^25p^5$ 126.904	**Xe** 54 $5s^25p^6$ 131.30
6	**Cs** 55 $6s^1$ 132.905	**Ba** 56 $6s^2$ 137.33	**La*** 57 $5d^16s^2$ 138.905	**Hf** 72 $4f^{14}5d^26s^2$ 178.49	**Ta** 73 $5d^36s^2$ 180.948	**W** 74 $5d^46s^2$ 183.85	**Re** 75 $5d^56s^2$ 186.207	**Os** 76 $5d^66s^2$ 190.2	**Ir** 77 $5d^76s^2$ 192.22	**Pt** 78 $5d^96s^1$ 195.09	**Au** 79 $5d^{10}6s^1$ 196.967	**Hg** 80 $5d^{10}6s^2$ 200.59	**Tl** 81 $6s^26p^1$ 204.37	**Pb** 82 $6s^26p^2$ 207.19	**Bi** 83 $6s^26p^3$ 208.980	**Po** 84 $6s^26p^4$ (209)	**At** 85 $6s^26p^5$ (210)	**Rn** 86 $6s^26p^6$ (222)
7	**Fr** 87 $7s^1$ (223)	**Ra** 88 $7s^2$ (226)	**Ac†** 89 $6d^17s^2$ (227)	**Rf** 104 (260)	**Ha** 105 (260)													

f Orbitals being filled

* Lanthanides ~ $4f^n5d^{0-1}6s^2$

Ce 58 $4f^15d^16s^2$ 140.12	**Pr** 59 $4f^35d^06s^2$ 140.907	**Nd** 60 $4f^45d^06s^2$ 144.24	**Pm** 61 $4f^55d^06s^2$ (145)	**Sm** 62 $4f^65d^06s^2$ 150.35	**Eu** 63 $4f^75d^06s^2$ 151.96	**Gd** 64 $4f^75d^16s^2$ 157.25	**Tb** 65 $4f^95d^06s^2$ 158.925	**Dy** 66 $4f^{10}5d^06s^2$ 162.50	**Ho** 67 $4f^{11}5d^06s^2$ 164.930	**Er** 68 $4f^{12}5d^06s^2$ 167.26	**Tm** 69 $4f^{13}5d^06s^2$ 168.934	**Yb** 70 $4f^{14}5d^06s^2$ 173.04	**Lu** 71 $4f^{14}5d^16s^2$ 174.97

† Actinides ~ $5f^n6d^{0-1}7s^2$

Th 90 $5f^06d^27s^2$ 232.038	**Pa** 91 $5f^26d^17s^2$ (231)	**U** 92 $5f^36d^17s^2$ 238.03	**Np** 93 $5f^46d^17s^2$ (237)	**Pu** 94 $5f^66d^07s^2$ (244)	**Am** 95 $5f^76d^07s^2$ (243)	**Cm** 96 $5f^76d^17s^2$ (247)	**Bk** 97 $5f^96d^07s^2$ (247)	**Cf** 98 $5f^{10}6d^07s^2$ (251)	**Es** 99 $5f^{11}6d^07s^2$ (254)	**Fm** 100 $5f^{12}6d^07s^2$ (257)	**Md** 101 $5f^{13}6d^07s^2$ (258)	**No** 102 $5f^{14}6d^07s^2$ (255)	**Lr** 103 $5f^{14}6d^17s^2$ (260)

190

APPENDIX 2: Answers To Problems

Self-Test I

1. **(a)** 480 Hz/8.0 ppm = 60 MHz (Section 5-2); radio-frequency (rf, Section 1-2); $E = hv = 3.98 \times 10^{-26}$ J (Section 1-2).

(b) $B_0 = 60.0$ MHz/42.58 MHz $T^{-1} = 1.41$ T (Section 2-2B); two, since $I = \frac{1}{2}$ (Section 2-1); $v_{precession} = v_{irradiation} = 60$ MHz (Section 2-2B).

(c) The "ringing," apparent in the signal at δ 0.00 ppm, indicates a continuous-wave spectrum (Section 3-4) scanned from left to right.

(d) The hydrogens of the reference compound TMS (Section 5-1A).

(e)

Signal	δ (ppm)	relative intensity
a	3.65	3.5
b	2.62	7.0
c	2.18	11.7
d	1.25	22

If there is a total of 12 hydrogen atoms in the molecule, signal **a** must represent one hydrogen, **b** two, **c** three, and **d** six.

(f) Because there are four signals, the 12 hydrogens must be apportioned among four sets. The only structure that will give this number of signals with the observed relative intensities is

(g) δ_a: exact calculation impossible, but the signal's position is within the limits normal for an alcohol hydroxyl group (δ 2–4 ppm). Also, the signal is quite broad, a characteristic of hydroxyl signals (Section 6-7A).

$$\delta_b = \delta(\text{methane}) + \Delta\delta\left(-C\underset{R}{\overset{O}{\diagup}}\right) + \Delta\delta(R) = 0.23 + 1.70 + 0.47 = 2.40 \text{ ppm}$$

$$\delta_c = \delta(\text{methane}) + \Delta\delta\left(-C\underset{R}{\overset{O}{\diagup}}\right) = 0.23 + 1.70 = 1.93 \text{ ppm}$$

$$\delta_d = \delta(\text{methane}) + \Delta\delta(R) = 0.23 + 0.47 = 0.70 \text{ ppm (see Section 6-3A)}$$

note: The extra deshielding of the hydrogens may be due in part to the $\Delta\delta(R)$ values that neglect the second oxygen and in part to the effects of hydrogen bonding (Section 6-7A):

2. **(a)** $Z = 6$; $N = 7$ (Section 2-1A).

(b) $v_0 = \dfrac{5000 \text{ Hz}}{200 \text{ ppm}} = 25.0$ MHz (Section 5-2).

(c) $I = \frac{1}{2}$ (Section 2-1); this tells us that carbon nuclei can adopt only two spin states when immersed in an external magnetic field because the number of possible spin states equals $2I + 1$ (Section 2-2).

(d) Virtually all ^{13}C spectra are acquired under PFT conditions (Section 3-3).

(e) The signal at δ 0.00 is due to the reference compound TMS; the signals at δ 76.0, 77.5, and 79.0 are due to the solvent $CDCl_3$ (Section 5-4A).

(f) δ 41.1, 51.6, 127.1, 128.6, 129.4, 134.4, and 171.6. The separation between the signals at δ 41.1 and 51.6 is

$$\Delta v \doteq (51.6 - 41.1 \text{ ppm})(25.0 \text{ Hz/ppm}) = 262.5 \text{ Hz}$$

(g) The structure that best fits the number, positions, and intensities of the signals is

Carbon	$\delta_{calculated}$ (ppm)	$\delta_{observed}$ (ppm)	Section
1	$\delta(CH_4) + \Delta\delta$ $-O-C\overset{O}{\underset{R}{\diagdown}} = -2.3 + 51 = 48.7$	51.6	7-2
2	range 165–175	171.6	7-6
3	$\delta(CH_4) + \Delta\delta\left(-C\overset{O}{\underset{OR}{\diagdown}}\right) + \Delta\delta(\text{phenyl}) = -2.3 + 20 + 23 = 40.7$	41.1	7-2
4	$\delta(\text{benzene}) + \Delta\delta(-CH_2CH_3, \alpha) = 128.5 + 15.6 = 144.1$	134.4	7-4
5	$\delta(\text{benzene}) + \Delta\delta(-CH_2CH_3, o) = 128.5 + (-0.5) = 128.0$	128.6	7-4
6	$\delta(\text{benzene}) + \Delta\delta(-CH_2CH_3, m) = 128.5 + 0.0 = 128.5$	129.4	7-4
7	$\delta(\text{benzene}) + \Delta\delta(-CH_2CH_3, p) = 128.5 + (-2.6) = 125.9$	127.1	7-4

note: Structures such as these fit the molecular formula but do not fit the chemical shift data as well:

3. (a) This is an easy one. The 5H signal at δ 7.28 ppm is characteristic of a phenyl (C_6H_5) ring attached to a nonpolar (carbon) substituent group (see Section 5-1A). Thus, the remaining atoms can be connected in only one way:

The predicted chemical shifts of the methylene hydrogens can be calculated as

$$\delta(CH_4) + \Delta\delta(OH) + \Delta\delta(Ph) = 0.23 + 2.56 + 1.85 = 4.64 \text{ ppm}$$

(b) The signal intensities suggest a CH_3 group and a CH_2 group. With only three carbons, there are only two possibilities with the required symmetry:

$$H_3C-CH_2-CCl_3 \quad \text{and} \quad H_3C-C(Cl)_2-CH_2Cl$$

$$\textbf{A} \qquad\qquad\qquad\qquad \textbf{B}$$

The chemical shift of the methylene group (δ 4.02 ppm) requires that it be attached directly to the deshielding chlorine substituent. Only structure **B** fits this requirement. The predicted chemical shifts can be calculated from the $\Delta\delta$ values for Cl (2.53 ppm, Table 6-1) and for

$-\overset{|}{\underset{|}{C}}-Cl$ ($\sim$0.5 ppm, Section 6-3B).

$$\delta_{CH_3} = \delta(CH_4) + 2\Delta\delta(CCl) = 0.23 + 2(0.5) = 1.23 \text{ ppm}$$

$$\delta_{CH_2} = \delta(CH_4) + \Delta\delta(Cl) + 2\Delta\delta(CCl) = 0.23 + 2.53 + 2(0.5) = 3.76 \text{ ppm}$$

Both of these calculated values are somewhat less than the observed values, perhaps indicating some special effects operating at the central carbon (see Example 6-7). In Chapter 8 you'll learn another way to have ruled out structure **A**.

(c) Because these two compounds have the same molecular formula but different structures, they are isomers. The close similarity of their ^{1}H spectra indicates that their structures are probably very similar. Each isomer has a 10H signal near δ 7.3 ppm, suggesting the presence of two equivalent phenyl rings attached to nonpolar substituents (see Problem 3a). The two remaining equivalent hydrogens near δ 4.0 ppm are in the correct range for a hydrogen on carbon connected to an oxygen and a phenyl, as in Problem 3a. There are only two possible structures:

where Ph represents a phenyl group. But which structure belongs to which stereoisomer? Recalling that nuclei located directly above or below an aromatic ring experience a special *shielding* effect (Section 6-6 and Example 6-18), we note that the trans isomer places the methine hydrogens in such a position:

Therefore, the first set of data, with the lower value of δ(CH), corresponds to the trans isomer; the second set, with the higher value of δ(CH), corresponds to the cis isomer.

(d) To interpret ^{13}C spectral data, remember that the relative intensity of each signal is *not* a simple function of the number of carbons represented by that signal. Here we have six signals for ten carbons, so there is some symmetry to the structure. The four signals from δ 114.9 to 154.7 suggest an aromatic ring with two different substituents para to each other. The four remaining carbons, represented by the signals at δ 31.5 and 33.6 ppm, are too far upfield to be attached to oxygen. Furthermore, their relative intensity suggests that the signal at 31.5 ppm probably represents more carbons (with more hydrogens) than the signal at 33.7 ppm. Here is a structure that fulfills all these criteria:

The predicted chemical shift of each carbon can be calculated as follows:

C	$\delta_{calculated}$ (ppm)				δ_{obs}	Reference
1	δ(propane, C_1) + $\Delta\delta$(int. CH_3, β) + $\Delta\delta$(int. phenyl, β) =					
	15.8	+ 8	+ 7	= 30.8	31.5	Tables 7-1, 7-2
2	δ(propane, C_2) + $\Delta\delta$(int. CH_3, α) + $\Delta\delta$(int. phenyl, α) =					
	16.3	+ 6	+ 17	= 39.3	33.7	Tables 7-1, 7-2

C	$\delta_{\text{calculated}}$ (ppm)	δ_{obs}	Reference
3	$\delta(\text{benzene}) + \Delta\delta(-\text{C}_2\text{H}_5, \alpha) + \Delta\delta(\text{OH}, p) =$		
	$128.5 \quad + \quad 15.6 \quad + \quad (-7.3) \quad = 136.8$	141.6	Table 7-4
4	$\delta(\text{benzene}) + \Delta\delta(-\text{C}_2\text{H}_5, o) + \Delta\delta(\text{OH}, m) =$		
	$128.5 \quad + \quad (-0.5) \quad + \quad 1.4 \quad = 129.4$	125.9	Table 7-4
5	$\delta(\text{benzene}) + \Delta\delta(-\text{C}_2\text{H}_5, m) + \Delta\delta(\text{OH}, o) =$		
	$128.5 \quad + \quad 0 \quad + \quad (-12.7) \quad = 115.8$	114.9	Table 7-4
6	$\delta(\text{benzene}) + \Delta\delta(-\text{C}_2\text{H}_5, p) + \Delta\delta(\text{OH}, \alpha) =$		
	$128.5 \quad + \quad (-2.6) \quad + \quad 26.9 \quad = 152.8$	154.7	Table 7-4

(e) Once again, three signals for five carbons indicates a symmetrical structure. The signals at δ 108.3 and 121.6 ppm are in the vinyl or aromatic range, while the δ 35.6 ppm signal could be a methyl carbon attached to a nitrogen. If you read Chapter 7 carefully, you may remember from the end of Section 7-4 some examples of heteroaromatic compounds. Note how closely the two downfield signals match those of pyrrole. A methyl group on the nitrogen would preserve the molecule's symmetry and would have a predicted chemical shift of $\delta(\text{CH}_4) + \Delta\delta(-\text{NR}_2, \alpha$, terminal) $= -2.3 + 42 = 39.7$ ppm ($\delta_{\text{obs}} = 35.6$ ppm). Thus, the correct structure is

(f) Here we go again: seven carbons and five signals; symmetry. There appear to be no aromatic or vinyl carbons, but what's this signal at δ 183 ppm? A possible carbonyl carbon. None of the other signals is downfield enough to represent a carbon attached to oxygen, though the one at δ 43.1 ppm looks like a carbon attached to a carbonyl carbon. The only way we can have a carbonyl group plus another oxygen, with *no* other carbons attached to that oxygen, is to have a carboxylic acid (Table 7-5):

When you try to write structures for the remaining six carbons and eleven hydrogens, you'll quickly discover that there must be a ring in the structure. Here are the four possibilities:

Of these, only the first one correlates with the chemical shift data:

C	$\delta_{\text{calculated}}$ (ppm)	δ_{obs}	Reference
1	range 170–185	183.0	Table 7-5
2	$\delta(\text{cyclohexane}) + \Delta\delta(-\text{CO}_2\text{H}, \alpha, \text{internal}) = 26.9 + 16 = 42.9$	43.1	Tables 7-1, 7-2
3	$\delta(\text{cyclohexane}) + \Delta\delta(-\text{CO}_2\text{H}, \beta, \text{internal}) = 26.9 + 2 = 28.9$	29.0	Tables 7-1, 7-2
4	$\delta(\text{cyclohexane}) + \Delta\delta(\text{int.} -\text{CO}_2\text{H}, \gamma, \text{internal}) = 26.9 + (-2) = 24.9$	25.9	Tables 7-1, 7-2
5	$\delta(\text{cyclohexane}) = 26.9$	25.5	Tables 7-1, 7-2

The last two values are too close to differentiate unambiguously.

4. (a) Because there's no molecular formula, this is a tough one. The ^{1}H spectrum exhibits only two signals (δ 1.41 and 1.90 ppm) in the ratio 3:1, respectively. The 3:1 hydrogen *ratio* doesn't necessarily mean there are only four hydrogens. There could be eight (in the ratio 6:2), or twelve (9:3), etc. The ^{13}C spectrum exhibits four signals (δ 22.3, 28.2, 79.8, and 170.0 ppm). The last of these most likely represents a carbonyl carbon, perhaps in an ester group. The presence of both a carbonyl carbon and a signal near δ 2.0 ppm in the ^{1}H spectrum suggests a methyl group attached to the carbonyl (as in Examples 6-4, 6-10, and 6-15):

$$\underset{\text{H}_3\text{C}-\overset{\displaystyle\text{O}}{\overset{\|}{\text{C}}}-}{}$$

This would require that the ^{1}H signal at δ 1.41 ppm represent *nine* equivalent hydrogens, as in a tertiary butyl group:

$$-\overset{\displaystyle\text{CH}_3}{\underset{\displaystyle\text{CH}_3}{\overset{|}{\underset{|}{\text{C}}}}}-\text{CH}_3$$

We cannot simply connect these two fragments for two reasons. First, the resulting structure would be a *ketone* (rather than an ester), and its carbonyl carbon would appear farther downfield (Table 7-5). Second, the resulting molecular weight (100) would be 16 units less than the value given. Both of these difficulties can be overcome, however, by including an oxygen atom:

$$\overset{\text{a}\ \text{H}_3\text{C}}{\underset{4\ \diagdown\text{H}_3\text{C}}{\text{H}_3\text{C}-\underset{|}{\overset{|}{\text{C}}}-\text{O}-\overset{\overset{\displaystyle\text{O}}{\|}}{\text{C}}-\text{CH}_3\ \text{b}}}$$

(b)

Atom	$\delta_{\text{calculated}}$ (ppm)	δ_{obs} (ppm)	Reference
C_1	$\delta(\text{CH}_4) + \Delta\delta\left(-\text{C}\diagup^{\text{O}}_{\diagdown\text{OR}}\right) = -2.3 + 20 = 17.7$	23.3	Section 7-2
C_2	range 165–175	170.0	Section 7-6
C_3	$\delta(\text{propane, } C_1) + \Delta\delta(\text{CH}_3, \beta) + \Delta\delta\left(-\text{O}-\text{C}\diagup^{\text{O}}_{\diagdown\text{R}}, \beta\right) = 15.8 + 8 + 5 = 28.8$	28.2	Section 7-2
C_4	$\delta(\text{propane, } C_2) + \Delta\delta(\text{CH}_3, \alpha) + \Delta\delta\left(-\text{O}-\text{C}\diagup^{\text{O}}_{\diagdown\text{R}}, \alpha\right) = 16.3 + 6 + 45 = 67.3$	79.8	Section 7-2
H_a	$\delta(\text{CH}_4) + \Delta\delta(\text{R}) = 0.23 + 0.47 = 0.70^*$	1.41	Section 6-2
H_b	$\delta(\text{CH}_4) + \Delta\delta\left(-\text{C}\diagup^{\text{O}}_{\diagdown\text{OR}}\right) = 0.23 + 1.55 = 1.78$	1.90	Section 6-2

* The $\Delta\delta$ value for R (Table 6-1) does not take into account the effect of the β oxygen, so it underestimates the deshielding effect of the R group.

Can you see how much more difficult it would be to deduce the structure without knowing the molecular weight of the compound? Fortunately, the techniques of mass spectrometry and elemental analysis routinely offer molecular weight and molecular formula information.

Self-Test II

1. (a) For a definition of "condensed form," see Section 8-2.

^{1}H: δ 1.99 (singlet, 3H), 7.3 (multiplet, 5H); the latter signal is designated a multiplet because its exact multiplicity cannot be determined by visual inspection.

^{13}C: δ 4.1, 80.1, 85.9, 124.4, 127.6, 128.3, 131.7

(b) The five-hydrogen multiplet at δ 7.3 ppm and the four carbon signals at δ 124–132 ppm suggest a monosubstituted aromatic ring, which would have both a C_2 axis and a plane of symmetry. The three-hydrogen singlet at δ 1.99 ppm indicates three equivalent hydrogens that are likely to be part of a methyl group. The two ^{13}C signals near δ 85 ppm are in the region for acetylenic carbons, and their intensities are low as would be expected for carbons bearing no hydrogens. The only structure that fulfills all these facts is

(c)

Nucleus	$\delta_{observed}$ (ppm)	$\delta_{calculated}$ (ppm)	Reference
$H_{1,2,3}$	7.2	~ 7.4	Example 6-16
H_4	1.99	δ(methane) + $\Delta\delta$(—C≡C—)	
		= 0.23 + 1.44 = 1.67	Table 6-1
C_1	127.6	δ(benzene) + $\Delta\delta$(—C≡C—, p)	
		= 128.5 + (−0.2) = 128.3	Table 7-4
C_2	128.3	δ(benzene) + $\Delta\delta$(—C≡C—, m)	
		= 128.5 + 0.4 = 128.9	Table 7-4
C_3	131.7	δ(benzene) + $\Delta\delta$(—C≡C—, o)	
		= 128.5 + 3.8 = 132.3	Table 7-4
C_4	4.1	δ(methane) + $\Delta\delta$(—C≡C—, α)	
		= −2.3 + 4.5 = 2.2	Table 7-2
C_5	124.4	δ(benzene) + $\Delta\delta$(—C≡C—, α)	
		= 128.5 + (−6.1) = 122.4	Table 7-4
C_6, C_7	85.9	range 70–90	Section 7-5

(d) The signal for C_4 will be split by the three methyl hydrogens into a *quartet* (Section 8-3). To find the value of the residual coupling constant, use Eq. (11-2) and the value of $^1J_{CH}$ for sp^3-hybridized carbons (Table 9-1):

$$J_r = \frac{2\pi J \, \Delta\nu}{\gamma B_2} = \frac{2\pi(125 \text{ Hz})(2000 \text{ Hz})}{(267.5 \times 10^6 \text{ rad T}^{-1}\text{ Hz})(2 \times 10^{-4} \text{ T})} = 29 \text{ Hz}$$

The signals for C_1, C_2, and C_3 will be split by one hydrogen each into *doublets*. The hybridization of carbons in an aromatic ring is sp^2, so use a $^1J_{CH}$ value of 156 Hz (Table 9-1):

$$J_r = \frac{2\pi J \, \Delta\nu}{\gamma B_2} = \frac{2\pi(156 \text{ Hz})(2000 \text{ Hz})}{(267.5 \times 10^6 \text{ rad T}^{-1}\text{ Hz})(2 \times 10^{-4} \text{ T})} = 37 \text{ Hz}$$

2. (a) ^{1}H: δ 0.95 (triplet, $J = 7$ Hz, 3H), 1.22 (doublet, $J = 7$ Hz, 6H), 1.70 (multiplet, $J = 7$ Hz, 2H), 2.18 (triplet, $J = 7$ Hz, 2H), 4.92 (multiplet, $J = 7$ Hz, 1H).

^{13}C: δ 13.7, 18.7, 22.0, 36.6, 67.1, 172.7

(b) First, consider the ^{1}H spectrum. The triplet at δ 0.95 ppm is likely to be due to a methyl group coupled to two equivalent hydrogens, as in CH_3—CH_2—. The six-hydrogen doublet (δ 1.22 ppm) suggests two *equivalent* methyl groups split by *one* neighboring hydrogen, as in $(CH_3)_2CH$— (an isopropyl group). The methine hydrogen of the isopropyl group is found as the expected septet at δ 4.92 ppm, indicating that the isopropyl group is attached to an oxygen, $-O-CH(CH_3)_2$. The two-hydrogen triplet at δ 2.18 ppm suggests a CH_2 group attached to a carbonyl and split by two hydrogens, i.e., $O=\overset{|}{C}-CH_2-CH_2-$. The presence of an ester carbonyl is confirmed by the ^{13}C signal at δ 172.7 ppm. Putting all this together in one structure gives

$$CH_3-CH_2-CH_2-\overset{\overset{O}{\|}}{C}-O-\overset{\overset{CH_3}{|}}{\underset{\underset{H}{|}}{C}}-CH_3$$

Note how the number and positions of the ^{13}C signals are consistent with this structure.

(c)

Nucleus	$\delta_{observed}$ (ppm)	$\delta_{calculated}$ (ppm)	Reference
H_1	0.95	δ(methane) + $\Delta\delta(CH_3)$ = 0.23 + 0.47 = 0.70	Table 6-1
H_2	1.70	δ(methane) + 2$\Delta\delta(CH_3)$* = 0.23 + 2(0.47) = 1.17	Table 6-1
H_3	2.18	δ(methane) + $\Delta\delta(CH_3)$ + $\Delta\delta(-CO_2R)$ = 0.23 + 0.47 + 1.55 = 2.25	Table 6-1
H_4	1.22	δ(methane) + $\Delta\delta(CH_3)$* = 0.23 + 0.47 = 0.70	Table 6-1
H_5	4.92	δ(methane) + 2$\Delta\delta(CH_3)$ + $\Delta\delta(-O_2CR)$ = 0.23 + 2(0.47) + 3.13 = 4.30	Table 6-1
C_1	13.7	δ(propane, C_1) + $\Delta\delta(-CO_2R, \gamma)$ = 15.8 + (-2) = 13.8	Table 7-1
C_2	18.7	δ(propane, C_2) + $\Delta\delta(-CO_2R, \beta)$ = 16.3 + 3 = 19.3	Tables 7-1, 7-2
C_3	36.6	δ(propane, C_1) + $\Delta\delta(-CO_2R, \alpha)$ = 15.8 + 20 = 35.8	Tables 7-1, 7-2
C_4	22.0	δ(propane, C_1) + $\Delta\delta(-O_2CR, \beta)$ = 15.8 + 5 = 20.8	Tables 7-1, 7-2
C_5	67.1	δ(propane, C_2) + $\Delta\delta(-O_2CR, \alpha)$ = 16.3 + 45 = 61.3	Tables 7-1, 7-2
C_6	172.7	range 165–175	Table 7-5

* Neglects the additional deshielding by the neighboring oxygen.

(d) A τ value of 8 ms corresponds to a J value of 125 Hz (1/8 ms = 125 Hz, Section 11-8). Such a J value is typical of a one-bond coupling between hydrogen and sp^3 carbon (Table 9-1). Figure 11-12 indicates that, with τ set to $1/J$, methylene and quaternary carbon signals will be positive, while methyl and methine carbon signals will be negative. Thus, the ^{13}C signals at δ 13.7 (C_1), 22.0 (C_4), and 67.1 (C_5) ppm will be negative; the other three will be positive.

3. (a) ^1H: δ 4.63 (singlet, 2H), 7.56 (doublet, J = 11 Hz, 2H), 8.19 (doublet, J = 11 Hz, 2H);

^{13}C: δ 44.6, 123.8, 129.8, 144.9, 147.5

(b) The two doublets in the ^1H spectrum and the four ^{13}C signals from δ 120 to 150 ppm strongly suggest a para-disubstituted aromatic ring. The remaining singlet in the ^1H spectrum is most likely a CH_2 (methylene) group attached to the ring. From the molecular formula, this leaves only $ClNO_2$, likely to occur as a Cl and an NO_2 (nitro) group. So there are two structures you might suggest:

$$O_2N-\langle\bigcirc\rangle-CH_2-Cl \qquad Cl-\langle\bigcirc\rangle-CH_2-NO_2$$

i   ii

The ^1H and ^{13}C chemical shifts of the methylene group readily distinguish between them.

Compound	δ_C	δ_H
i	$\delta(\text{methane}) + \Delta\delta(\text{Cl}, \alpha) + \Delta\delta(\text{phenyl}, \alpha)$ $= -2.3 + 31 + 23 = 52$	$\delta(\text{methane}) + \Delta\delta(\text{Cl}) + \Delta\delta(\text{phenyl})$ $= 0.23 + 2.53 + 1.85 = 4.61$
ii	$\delta(\text{methane}) + \Delta\delta(\text{NO}_2, \alpha) + \Delta\delta(\text{phenyl}, \alpha)$ $= -2.3 + 63 + 23 = 84$	$\delta(\text{methane}) + \Delta\delta(\text{NO}_2) + \Delta\delta(\text{phenyl})$ $= 0.23 + 3.80 + 1.85 = 5.88$

Clearly, structure **i** is a much better fit than **ii**.

(c) In addition to the methylene signals, all the other ^1H and ^{13}C signals are well correlated with structure **i**.

Nucleus	$\delta_{observed}$ (ppm)	$\delta_{calculated}$ (ppm)	Reference
H_2	7.56	$\delta(\text{benzene}) + \Delta\delta(\text{CH}_3, o) + \Delta\delta(\text{NO}_2, m)$ $= 7.27 + (-0.10) + 0.30 = 7.47$	Table 6-4
H_3	8.19	$\delta(\text{benzene}) + \Delta\delta(\text{CH}_3, m) + \Delta\delta(\text{NO}_2, o)$ $= 7.27 + (-0.10) + 0.97 = 8.14$	Table 6-4
C_2	129.8	$\delta(\text{benzene}) + \Delta\delta(\text{CH}_3, o) + \Delta\delta(\text{NO}_2, m)$ $= 128.5 + 0.7 + 0.9 = 130.1$	Table 7-4
C_3	123.8	$\delta(\text{benzene}) + \Delta\delta(\text{CH}_3, m) + \Delta\delta(\text{NO}_2, o)$ $= 128.5 + (-0.1) + (-4.8) = 123.6$	Table 7-4
C_4	144.9	$\delta(\text{benzene}) + \Delta\delta(\text{CH}_3, \alpha) + \Delta\delta(\text{NO}_2, p)$ $= 128.5 + 8.9 + 5.8 = 143.2$	Table 7-4
C_5	147.5	$\delta(\text{benzene}) + \Delta\delta(\text{CH}_3, p) + \Delta\delta(\text{NO}_2, \alpha)$ $= 128.5 + (-2.9) + 20 = 145.6$	Table 7-4

(d) See Section 11-9B. The t_1 value of 4 ms ($= 1/2J$) correlates signals that share a heteronuclear J value of 125 Hz. This value is typical for one-bond coupling between a hydrogen and an sp^3 carbon (Table 9-1). Figure A2-1 shows only one signal at the intersection of the signals for C_1 and H_1, the only pair that shares a 125-Hz coupling. Setting t_1 to 3.1 ms tests for $J = 160$ Hz couplings, typical for hydrogens attached to vinyl or aromatic carbons (Table 9-1). Figure A2-2, therefore, shows two signals, one at the intersection of the C_2 and H_2 signals, the other at the intersection of the C_3 and H_3 signals.

4. (a) Begin by estimating the magnitude of the three possible C–D coupling constants $^1J_{CD}$, $^2J_{CD}$, and $^3J_{CD}$. Use the proportionality between J_{XY} and $\gamma_X\gamma_Y$ (Section 9-2) to estimate J_{CD} values from the J_{CH} values in Tables 9-1, 9-3, and 9-4. Thus,

$$^1J_{CD} = \frac{\gamma_D}{\gamma_H}(^1J_{CH}) = \frac{41.064}{267.512}(125 \text{ Hz}) = 19 \text{ Hz}$$

Likewise, $$^2J_{CD} \approx {}^3J_{CD} = \frac{\gamma_D}{\gamma_H}(5 \text{ Hz}) < 1 \text{ Hz}$$

Therefore, the only coupling of any significance is ^1J. The carbonyl carbon should appear as a weak singlet at about δ 210 ppm (Table 7-5). The remaining two carbons are symmetry-equivalent and will appear at about δ 28 ppm (Table 7-2). From Eq. (8-3), $L = \prod_i (2n_i I_i + 1)$, the

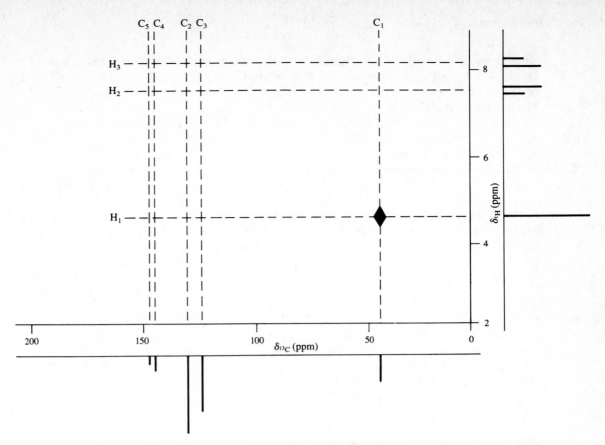

FIGURE A2-1 Idealized HETCOR spectrum of $ClCH_2C_6H_4NO_2$ with $t_1 = 4$ ms.

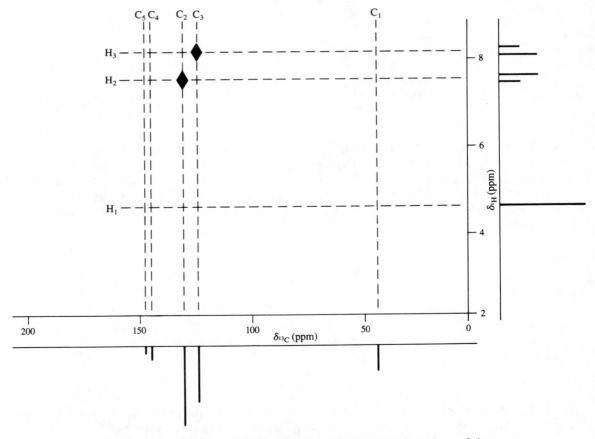

FIGURE A2-2 Idealized HETCOR spectrum of $ClCH_2C_6H_4NO_2$ with $t_1 = 3.1$ ms.

latter signal will be split by the three ($n_i = 3$) deuterium nuclei ($I = 1$) into seven lines ($L = 2 \cdot 3 \cdot 1 + 1 = 7$). The intensity ratio of the lines in this multiplet is 1:3:6:7:6:3:1, which can be deduced from the number of spin combinations that give rise to each M value (Section 8-6B). The line spacings will all be 19 Hz. (Review Example 8-16b).

(b) This spectrum will consist of one signal centered at δ 66 ppm (Table 7-2) that is split into a doublet (by the fluorine, $^1J_{CF} = 280–350$ Hz) of quartets (by the three equivalent hydrogens, $^1J_{CH} = 125$ Hz); see Table 9-2.

5. Typical vinyl hydrogens usually appear around δ 5.3 ppm (Section 6-4). But hydrogens attached to aromatic rings are further deshielded to about δ 7.3 ppm by delocalization of the electrons in the double bonds (Section 6-6). The ^1H chemical shift of cyclooctatetraene is much closer to that of "normal" vinyl hydrogens, indicating little, if any, "aromatic" delocalization of the electrons in its double bonds. In fact, unlike the planar molecule benzene (Example 4-3), cyclooctatetraene is tub-shaped, making interaction of the double bonds essentially impossible:

6. (a) From Table 7-2 you can see that the deshielding effect of a ketone carbonyl falls off monotonically as the number of intervening bonds increases. Therefore, the three methylene carbon signals can be assigned by their proximity to the carbonyl:

C_1 211.1
C_2 41.9
C_3 27.2
C_4 25.1

(b) We'll include only the effects of alpha substituents.

Carbon	$\delta_{calculated}$ (ppm)	$\delta_{observed}$ (ppm)	Reference
C=O	δ(cyclohexanone, C_1) = 211.2	217.6	part (a)
C—OH	δ(cyclohexanone, C_2) + $\Delta\delta$(OH, α) + $\Delta\delta$(CH$_3$, α)		
	= 41.9 + 41 + 6 = 89	83.5	Table 7-2
C—H	δ(cyclohexanone, C_2) + $\Delta\delta$(CH$_3$, α)		
	= 41.9 + 8 = 50	52.5	Table 7-2

(c) The molecule has a C_2 axis (Section 4-1) through its center and perpendicular to the ring. Thus, each methylene group is symmetry-equivalent to one other methylene group. Therefore, the fourteen CH$_2$ groups are divided among seven sets. Methylenes 1 and 7 are the most deshielded (by the OH and C=O groups) and appear as the two signals at δ 29.1 and 26.1 ppm, respectively.

The remaining five CH$_2$ signals are not resolved and appear together as the signal farthest upfield.

12-6

7. (a) The two terminal vinyl hydrogens (H$_a$) are equivalent (by virtue of a symmetry plane in the page) and experience long-range coupling to H$_b$ ($^4J = 6$ Hz, Table 9-5).

A

Thus the H_a signal, which should appear around δ 5.3 ppm (Section 6-4), will be split by H_b into a *doublet*. H_c will be split by the two H_a's into a triplet centered near δ 5.7 ppm (Table 6-3). The nine equivalent hydrogens of the tert-butyl group are not coupled to any other nuclei, so they give rise to a 9H singlet near δ 1.0 ppm (Example 6-10).

(b) Although the 9H singlet is exactly where you predicted, the vinyl hydrogens are about 0.6 ppm more shielded than expected. This is a characteristic of the $C=C=C$ linkage. Note, however, that the "doublet" at δ 4.7 ppm and the "triplet" at δ 5.1 ppm constitute an A_2B system that is rendered complex by second-order effects (Section 9-9). This is because $\Delta v/J$ is relatively small (24 Hz/6 Hz = 4). The two small signals flanking the 9H singlet are spinning side bands (Section 3-2C). The spacing between them (72 Hz) is too small for them to be ^{13}C satellites (Section 8-6A).[1]

(c) Only the CH_3 hydrogens are equivalent. The asymmetric carbon (*) renders the CH_2 hydrogens (H_6 and H_7) and the $=CH_2$ hydrogens (H_1 and H_2) *diastereotopic* (Section 4-3).[2]

B

(d) The half-width ($v_{1/2}$) of the O—H resonance will be broadened by hydrogen exchange (Section 6-7A), mainly because this exchange decreases the effective spin–spin relaxation time (T_2^*), which directly determines half-width (Sections 3-4A and 12-2B). In the case of an N—H signal there is an additional effect operating. The electric quadrupole of nitrogen also decreases T_2^*.

(e) Since there are eight sets of hydrogen, you'll need eight δ values. We can estimate these as follows:

H	$\delta_{calculated}$ (ppm)	Reference
1	4.7	part (b)
2	4.7	part (b)
3	5.1	part (b)
4	δ(methane) $+ 2\Delta\delta$(R) $+ \Delta\delta$(—C$=$C—) $= 0.23 + 2(0.47) + 1.32 = 2.5$	Table 6-1
5	δ(methane) $+ \Delta\delta$(R) $= 0.23 + 0.47 = 0.70$	Table 6-1
6	δ(methane) $+ \Delta\delta$(R) $+ \Delta\delta$(OH) $= 0.23 + 0.47 + 2.56 = 3.26$	Table 6-1
7	δ(methane) $+ \Delta\delta$(R) $+ \Delta\delta$(OH) $= 0.23 + 0.47 + 2.56 = 3.26$	Table 6-1
8	range 2–4	Section 6-7A

Note that H_1 and H_2 are expected to be very similar, as are H_6 and H_7. Neglecting coupling to the OH hydrogen (because of exchange), the following homonuclear H—H couplings will be > 1 Hz:

H_a	H_b	n^a	$^nJ_{ab}$	Reference
1	2	2	2.5	Table 9-3
1	3	4	6	Table 9-5
2	3	4	6	Table 9-5
1	4	5	3	Table 9-5
2	4	5	3	Table 9-5
3	4	3	7	Section 9-5
4	5	3	7	Section 9-5
4	6	3	7	Section 9-5
4	7	3	7	Section 9-5

^{a} Number of intervening bonds.

Note that H_4 is coupled to every other hydrogen in the molecule except H_8!

(f)

Signal	Multiplicity	Assignment
1.02	d	H_5
2.3	m	H_4
3.2	dd(?)	H_6, H_7
4.1	s	H_8
4.7	dd	H_1, H_2
5.1	dt	H_3

From Eq. (8-2), with $n_1 = n_2 = n_3 = n_6 = n_7 = 1$ and $n_5 = 3$,

$$L = \prod_i (2n_i I_i + 1) = \prod_i [2n_i(\tfrac{1}{2}) + 1] = \prod_i (n_i + 1)$$

$$= (1 + 1)(1 + 1)(1 + 1)(1 + 1)(1 + 1)(3 + 1) = 128$$

(g) Irradiation of the H_4 signal will decouple it from all other nuclei [see part (e)]. The decoupled spectrum will consist of the following signals: δ 1.02 (singlet), 3.2 (AB quartet because of asymmetric center), 4.1 (singlet), 4.7 (doublet), 5.1 (triplet). The last two of these will resemble the A_2B pattern in Problem 7b.

(h) The ^{13}C signal from the middle carbon in a C=C=C linkage is usually weak and occurs far downfield, in the carbonyl region (δ 200–215 ppm, Section 7-6).

8. In structure **C** the two "outer" (meta to the $CONH_2$) OCH_3 groups are symmetry-equivalent and distinct from the "central" one (para to the $CONH_2$ group). At $+38°$ C the same is true of compound **D**. Yet, as the temperature is lowered, it is clear that the two "outer" OCH_3 groups in **D** become nonequivalent. This indicates that, although they are inherently nonequivalent, there is some type of process that averages them at the higher temperature. This process is restricted rotation around the bond between the single ring and fused ring system in **D**, caused by crowding of the two ring systems:[3]

The reason why only one signal is apparent at $-23°$ C is that the two exchanging peaks have coalesced into one very broad one (Section 12-2), obscured by baseline noise.

9. Recall from Section 11-9 that in a COSY spectrum both axes correspond to ^1H chemical shifts. The normal spectrum (in contour) appears along the diagonal, while off-diagonal signals correspond to signals that are coupled. Of the 20 hydrogens in the structure, only those on the methyl group are equivalent, so we expect to find 18 signals in the ^{1}H spectrum. Of these, 17 signals (with their apparent multiplicity) can be discerned: δ 1.280 (d, 3H); 1.674 (m, 1H); 1.756 (m, 2H); 2.129 (dt, 1H); 2.466 (td, 1H); 2.623 (q, 1H); 2.777 (dd, 1H); 2.920 (d, 1H); 3.484 (td, 1H); 3.731 (d, 1H); 4.294 (q, 1H); 5.562 (br. s, 1H); 6.890 (d, 1H); 7.026 (t, 1H); 7.198 (t, 1H); 7.385 (d, 1H); and 8.149 (s, 1H) ppm.

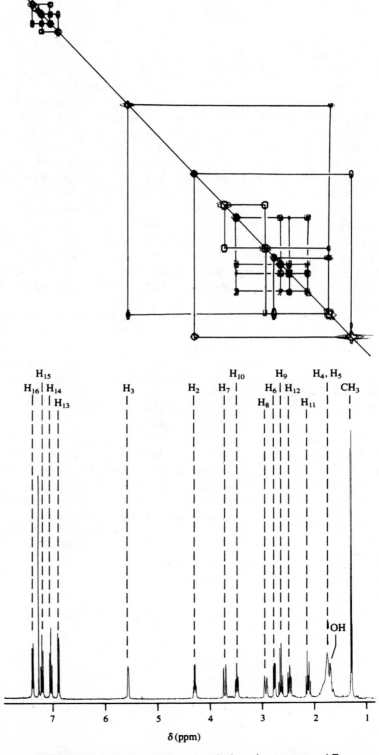

FIGURE A2-3 H—H coupling correlations for compound **E**.

The structure is labeled with hydrogen positions H_{10}, H_9, H_7, H_8, OH, H_{16}, H_{11}, H_{12}, N, CH_3, H_{14}, H_2, H_3, H_{15}, H_6, H_{13}, H, N, O, H_4, H_5.

E

You can begin by assigning the doublet at δ 1.280 ppm to the methyl group and the singlet at δ 8.149 ppm to the N—H (Section 6-7B). Next, draw a diagonal on the COSY spectrum in Figure II-10 (see Figure A2-3) and note that each signal in the spectrum below the COSY plot has a counterpart on the diagonal. Notice also that the COSY plot is symmetrical about the diagonal. From the center of the COSY methyl signal (at the lower right corner of the COSY spectrum), we trace faint lines horizontally and vertically, looking for intersecting signals. There is only one, the signal at δ 4.294 ppm. By looking at structure E, you can assign the latter signal to H_2. Show this correlation by drawing the indicated correlation "box."

From Section 6-4, the signal at δ 5.562 ppm must correspond to the vinyl hydrogen H_3. It is correlated with (and hence coupled to) the 2H signal at δ 1.756 ppm, which therefore represents H_4 and H_5 (accidental equivalence). These two hydrogens are further correlated with the signals at δ 2.777 and δ 2.920. Of these, only the signal at δ 2.920 is further coupled (to the signal at δ 3.731). This means that the signal at δ 2.777 corresponds to H_6. The signals at δ 2.920 and 3.731 ppm are due to H_7 and H_8, though the H_4 (or H_5) coupling to one of them must result from a "W"-type interaction through space (Section 9-6). The four remaining methylene hydrogens are all coupled to each other. H_9 and H_{10} should be more deshielded than H_{11} and H_{12}, owing to the proximity of the nitrogen. So, we can assign the signals at δ 2.129 and 2.466 to H_{11} and H_{12}, and the signals at δ 2.623 and 3.484 to H_9 and H_{10}.

Comparing the chemical shifts of the aromatic hydrogens to the values in Section 6-6 (Table 6-4), we can assign the signals as follows: δ 6.890, H_{13}; 7.026, H_{14}; 7.198, H_{15}; 7.385, H_{16}. Notice how only ortho (three-bond) couplings between these four hydrogens are in evidence.

This leaves the O—H hydrogen, which is the signal at δ 1.674. Thus, the COSY spectrum confirms all the expected couplings and makes the assignments of signals, even in a structure as complex as this one, relatively straightforward.

10. First, the magnetogyric ratio of Cl (Table 2-1) is small, and this affects the magnitude of J. But more importantly, a chlorine nucleus (as well as any nucleus with $I \geq 1$) has an electric quadrupole moment because of the nonspherical distribution of charge in its nucleus. This fact causes chlorine nuclei to undergo spin–lattice relaxation extremely rapidly on the nmr time scale. This rapid interconversion and equilibration among its various nuclear spin states effectively decouples the chlorine from spin–spin interactions with neighboring nuclei. However, the esr time scale is nearly a thousand times faster than the nmr time scale (Section 10-4), so the relaxation of the chlorine nuclei is not fast enough to decouple them from neighboring unpaired electrons.

REFERENCES

1. R. S. Macomber and K. C. Lilje, *J. Org. Chem.* **39**, 3600 (1974).

2. R. S. Macomber, *J. Org. Chem.* **36**, 999 (1971).

3. J. C. Schmidt, H. D. Benson, R. S. Macomber, B. Weiner, and H. Zimmer, *J. Org. Chem.* **42**, 2003 (1977).

APPENDIX 3

Compound Index

Index of ^{13}C Spectra

Index of ^{1}H Spectra

SUBJECT INDEX

α position, 73
AB quartet, 131
AB spin system, 128–131
AX spin system, 128–131
Absolute value, 122
Absorption mode, 136
Absorption, enhanced, net A effect, 139
Acetylenic carbon, 81
Acetylenic hydrogen, 62
Acid halide, 82
Acidity order, 66–67
Acquisition time, 28
Additivity principle for chemical shifts, 56–57
Alcohol, 66
Aldehyde, 82
Aldehydic hydrogen, 62
Allene, 84
Allowed transition, 108
Amide, 68, 82
Amine, 68, 75
Ammonium salt, 68
Amplitude, 1
Anhydride, 82
Aniline, 68
Anisotropy
    magnetic, 62–63
    chemical-shift, 178
APT, see Attached proton test
Aromatic carbon, 78–81
Aromatic compound, 63–66
Asymmetric center, 39
Atom, structure of, 6–9
Atomic mass, 6
Attached proton test, 160, 162
Axial position, 116
Axis of symmetry, 33–35
    local, 37
    rotational, 33

β position, 73
Beat pattern, 149
Boltzmann distribution, 12–13
Bond, see Chemical bond
Bond dissociation energy, 133
Broadband radiation, 149
Binomial distribution, 96

^{13}C nmr, 47–50, 72–87
C_n, see Axis of symmetry
Calculated spectra, 126–128
Carbonyl group, 62
    chemical shift of, 83
Carboxylate ester, 82
Carboxylic acid, 66, 82
CAT
    computer-averaged transients, 24
    computerized axial tomography, 180
Chemical bond, 111, 133–134
Chemical equivalence, 35

Chemical processes, 177–178
Chemical shift, 45–46
    anisotropy, 178
    reagent, 169–172
    δ scale, 45
    τ scale, 45–46
Chemical shift correlations
    base values for, 58
    carbon, 72–87
    hydrogen, 53–71
Chiral atom, 39
CIDNP, 139–146
Cis relationship, 61, 117
Coalescence, 174
Coil
    irradiating, 148
    receiving, 15, 17, 148
    sampling, see receiving
    sensing, see receiving
    surface, 180
    transmitting, 17, 148
Condensed form, 93
Conformation, 37
Conjugation, 78
Continuous wave, 20–22
Contour plot, 163
COSY, 166–167
Coupling (spin–spin), 93
    constant (J), 92, 103–132
        long-range, 118–119
        one-bond, 112–114
        three-bond (vicinal), 115–118
        two-bond (geminal), 114–115
    heteronuclear, 100–103
    homonuclear, 93–94
    hyperfine, 136–138
CPMAS, 178
Cross polarization, 151
Cyano group, 82
Cyclic molecule, 57–58

δ scale, 45
Decoupling, 50, 100, 149–150, 152–154
    gated, 158–159
    heteronuclear, 149–150
    homonuclear, 153–154
    inverse gated, 159
    off-resonance, 152–153
    proton, 105
    selective, 156–157
Degenerate rearrangement, 178
Degenerate transitions, 108
Delay time, 26, 28
Delocalization of electrons, see Resonance
DEPT, 162
Deshielding, 53
Detection, 159
Diamagnetic species, 133
Diastereomer, 39
Diastereotopic nuclei, 39, 125

Dihedral angle, 115
Dispersion mode, 136
Double resonance techniques, 148–157
Doublet, 92
Downfield, 42
Dwell time, 25
Dynamic processes, 172–178

Electromagnetic radiation, 1
Electromagnetic spectrum, 2
Electronegativity, 60
Electrons, 6, 7, 133–136
Electron spin, 7
Emission, stimulated, 12
    net E effect, 140
Enantiomer, 38
Enantiotopic nuclei, 38
Energy, spin state, 9
Energy levels, nuclear, 12–15
Epr, see Esr
Equatorial position, 116
Equivalence
    accidental, 39
    chemical (shift), 35
    magnetic, 97
    symmetry, 35
Escape products, 143
Esr, 133–139
Ethoxy, 104, 122
Ethyl radical, 145
Evolution time, 159
Exchange, 67, 172–175
    deuterium, 67–68
    hydrogen, 67
Exchange ratio, 173

Fast exchange limit, 173–175
FID, see Free induction decay
Field
    irradiating, 148
    observing, 148
    sweep, 21–23
First-order coupling, 91–107
Fluxional molecules, 177
Forbidden transition, 108
Formate, 83
Fourier transformation, 27
Free electron density, 137
Free induction decay, 25
Free radical, 133
Frequency, 1
    irradiating, 148
    Larmor, 11–12
    linear precession, 11
    observing, 148
    sweep, 20–21

γ, see Magnetogyric ratio
γ position, 73
Gated decoupling, 158–159
Geminal relationship, 58